RCM
GUIDEBOOK

Building a Reliable Plant Maintenance Program

RCM GUIDEBOOK

Building a Reliable Plant Maintenance Program

Jim August, P.E.

PennWell®

PennWell Corporation
1421 South Sheridan Road
Tulsa, Oklahoma 74112-6600 USA

800.752.9764
+1.918.831.9421
sales@pennwell.com
www.pennwellbooks.com
www.pennwell.com

Managing Editor: Marla Patterson
Production Editor: Sue Rhodes Dodd
Cover design: Charles Thomas
Book design: Robin Remaley

Library of Congress Cataloging-in-Publication Data Available on Request

August, J. K.
RCM Guidebook: Building a Reliable Plant Maintenance Program
p. cm.
q. cm
Includes index
ISBN 1-59370-007-5

Printed in the United States of America

1 2 3 4 5 08 07 06 05 04

Contents

List of Figures

List of Tables

Preface

I struggled for several years before starting this effort. Why another book about applied reliability-centered maintenance (RCM)? Several good ones are available. Why write a third? Is there a market for another? Would it serve any useful purpose? After I struggled with the idea for a year, I concluded that another RCM book *was* in order. The others target engineers; something briefer—something for a more general audience—was needed. Something more "nuts and bolts," step-by-step, guide-like, with key insights.

Users need better guidance on practical RCM applications. Most engineers grasp the RCM process theory quickly but struggle for years before producing useful periodic maintenance (PM) plans efficiently. Part of this is maintenance illiteracy common among engineers; unless they've *worked* in maintenance, they don't understand the processes or culture. Process, technical, writing, people, and craft skills are needed to work effectively with maintenance.

I believe that a great degree of the challenge performing RCM goes back to the design process. Complex process and facility design is fascinating. Design incorporates culture; design builds from experience. Design makes assumptions, which are embedded implicitly in equipment selection, redundancies, and instrumentation package provisions. Latent design factors influence facility operational outcomes for virtually all a facility's operating life.

So this book *expands* upon available RCM literature. We hope to provide readers with a greater practical understanding of how wonderful facilities evolved as designs and how these designs influence maintenance options.

I would like to thank my charming wife Cindy Sue and children, Gregory and Tom. Long absences and hours generated these ideas. You learn by doing; book learning never compensates for experience.

Reducing a process to software code forces you to intuitively learn that process. My software coding partner and advisor, Krishna "Devan" Vasudevan, converted many abstractions to data models that capture our practical RCM experience in software. Devan and I have *lived* RCM process logic coding—testing that code, presenting applications to users, listening to their remarks, and then revising logic and formats to resolve their objections and capture their ideas. We put more time in this effort than we would care to acknowledge.

Software applications provide an acid test. We create elegant software from profound process awareness. Users, their software interactions, our other software applications, and other legacy software provide many reference comparisons. Useful insights guide us to improving design. Well-developed RCM software provides users insight. Our goal is to develop elegant, insightful software. When software is clear, users are happy, clients are productive, and theoretical concepts spring to life as practical information faster and easier.

My final acknowledgment is for those who have supported me—my peers and our company, as we struggled with our own learning. It's unusual to both understand theory well and to have practical experience. Most people settle comfortably into one world or the other—design or operations. Both skills combined develop exceptional RCM-based maintenance programs. Both help improve design.

Special clients to be acknowledged include Steve Coppock and Brain Ramey of Arizona Public Service's Palo Verde Nuclear Generating Station and Pete Simonich of PPL Montana's Colstrip Generating Station. Personal friends and confidants such as Mike Blossom, Scheduler at Xcel Energy's Arapahoe Station; Frank Novachek, Special Projects Manager, also of Xcel; our advisors—Earl Hill of LOMA, and Clair Schwan and their peers—helped with practical support. Finally, professional organizations—the ANS Utility Working Conference, American Society of Mechanical Engineers (ASME) Reliability Committee, and Society of Maintenance and Reliability Professionals provided open forums for practical reliability discussions that have proven invaluable. We learn by interacting with others. Through these groups I have gained professional friends and contacts that provide a wide support base for any technical discussion that arises. Their ideas are embedded here.

The author would like to especially thank Core, Inc. and Asset Works, Inc. for their permission using trim (RCMtrim) and PowerFM, respectively, and for permission to use the RCM and CMMS/EAMS displays provided to illustrate the many technical discussions.

1 Introduction

This book provides quick, simple reference material on practical RCM. In addition to maintenance professionals, the book will be useful to nonmaintenance managers and engineers or anyone who desires a broader overview of maintenance performance theory from a simple, nontechnical perspective.

What is RCM?

RCM is a maintenance plan development process. RCM was first noted in a 1978 publication sponsored by the U.S. Department of Defense. That work documented a process developed through more than 20 years of commercial-aviation experience that demonstrated success at over-achieving airline operation, reliability, and safety goals. Participants included the government—the Federal Aviation Association, the airline industry, the Airline Transport Association, individual airlines—especially United Airlines, its employees, and suppliers—and, especially, Boeing. Air travelers, as well as the general public at large, are the primary beneficiaries.

RCM focuses on two words: reliability and maintenance. While most people are plausibly comfortable with *maintenance,* the term *reliability* introduces lesser-appreciated meanings and contexts. Risk, probability, consequences, local effects, secondary interactions—these reliability ideas place most people on unsteady ground (*see* Fig. 1–1).

Fig. 1–1 Plant Active Trouble Reports: Morning Work List

Equipment in systems interacts; equipment in systems comprises industrial facilities, and systems provide functionality. Some systems create product or output directly; others provide support. Some systems provide *public health and safety* protection (e.g., emissions removal systems for coal-fired power plants); others monitor the status of intangible process elements.

System development

For systems, RCM identifies functions that matter, equipment providing those functions; and it classifies equipment in context. It answers the question, "Why does that function matter?" Although individuals know pieces of the puzzle, an organizational awareness requires collective insight to develop. Often, a system-integrated understanding has never fully developed.

In start up, an architect-engineer (AE) provides system design documents. Plant operating culture 20 years ago didn't stress understanding *why* systems were designed as they were, or what equipment provided as system functional utility. Understanding the reasons designs include various functions is more important than ever today because only by these reasons, which explain the basis for all equipment roles in a system, can an organization maximize production while maintaining safety and cost. These determine the first five classic RCM steps. Four are system-level steps; the fifth is an equipment-level step. The system-level steps are

1. Select the system
2. Pick system boundaries
3. Identify system functions
4. Express system functions as functional failures

The equipment step is

5. Develop component failures with Failure Modes & Effects Analysis (FMEA)

System implies connectivity, which, in turn, implies an equipment breakdown structure. An equipment list should be available to define the system. AE process and instrumentation drawings (P&IDs) define systems, establish boundaries, identify subsystems, and provide system-oriented material (references to design requirements, control, functional block diagrams, and other descriptions and drawings). P&IDs support physical layout and hardware design, modeling systems. Equipment breakdown structures in the form of engineering design equipment lists constitute equipment information tables. These are loaded into computerized maintenance management systems/equipment asset management systems (CMMS/EAMS). Modern plant maintenance work is controlled and tracked using these computerized applications.

The equipment list allows RCM to identify the system's equipment that *matters*—those with direct failure potential. Direct failures directly affect needed system functions. Equipment lists help investigate failures, classifying equipment failure modes by dominance (based on occurrence frequency), and determining exposure risk. With dominant failures known, appropriate preventive maintenance (PM) tasks can be selected by equipment type and way of failing (the failure mode).

The final two classic RCM steps are

6. Identify equipment failure exposure risk with the FMEA
7. Select tasks that cost-effectively manage and prevent failures

Somehow, RCM acquired a seven-step methodology. Looking at competing seven-step methods, one finds they differ substantially. There are *top down* and *bottom up* constructions.

Final effort adjusts selected task performance intervals based upon failure time-dependence characteristic (e.g., aging), and packages results. This latter process, called *task blocking* in original RCM development, creates cost-effective task packages that choreograph maintenance. Simply releasing tasks as individual work orders for work is ineffective, as demonstrated in early Maintenance Information Systems (MIS), which represented 1980s' computerized PM systems implementation. The systems lacked PM task blocking and re-blocking capability; they were consequently ineffective. Users did not recognize PM task grouping as a conscious work efficiency development step.

The seven-step process provides a succinct RCM outline. Steps are omitted or condensed in a seven-step summary. RCM analysts make RCM look simple, but common pitfalls made by new analysts are ingrained in the veteran analyst's psyche as hard-won experience. Others don't have bitterly learned experience under their belts. A goal here is to provide new developers with insights so they don't make the same errors commonly encountered developing RCM-based maintenance programs.

SAE JA-1011—the evaluation criteria for RCM processes—provides a second seven-step process. The general steps follow a similar outline as identified in the system- and equipment-level RCM steps:

- Identify functions associated with performance.
- What ways can the functions fail?
- What are the causes for function failure?
- What are the failure effects?
- How does each failure matter?
- What scheduled maintenance should be done to control failure?
- What are defaults if no suitable task can be found?

JA-1011 RCM flow summarizes traditional RCM. System functions and function failures initiate analysis at the highest level. Focus shifts to function failure causes—component failures; their effects; and their classification in safety, operational, and cost terms. Finally, selecting the scheduled maintenance tasks completes the sequence, identifying a default activity where no appropriate tasks can be found. The JA-1011 standard emphasizes component and failure mode details. Different emphasis sheds light on finer points that different experts and processes introduce, partly reflecting their specialty. Basic RCM proceeds from system definition and losses to supporting components providing functionality and their failure modes. Emphasis shifts to the modes, effects, and classification of failure modes.

Failure description is an art, with terminology and interpretations that depend on the failure location in the system-component-part hierarchy level. Component failure modes cause system failure. A mode output becomes cause input for the next equipment level up in the chain. A failure mode is *how* something fails; a cause is *why*. With a *black box* interpretation, a failure mode provides evidence of failure as lost output or service (*see* Fig. 1–2). A cause is the corresponding lost input or internal fault that prevents the output from being generated.

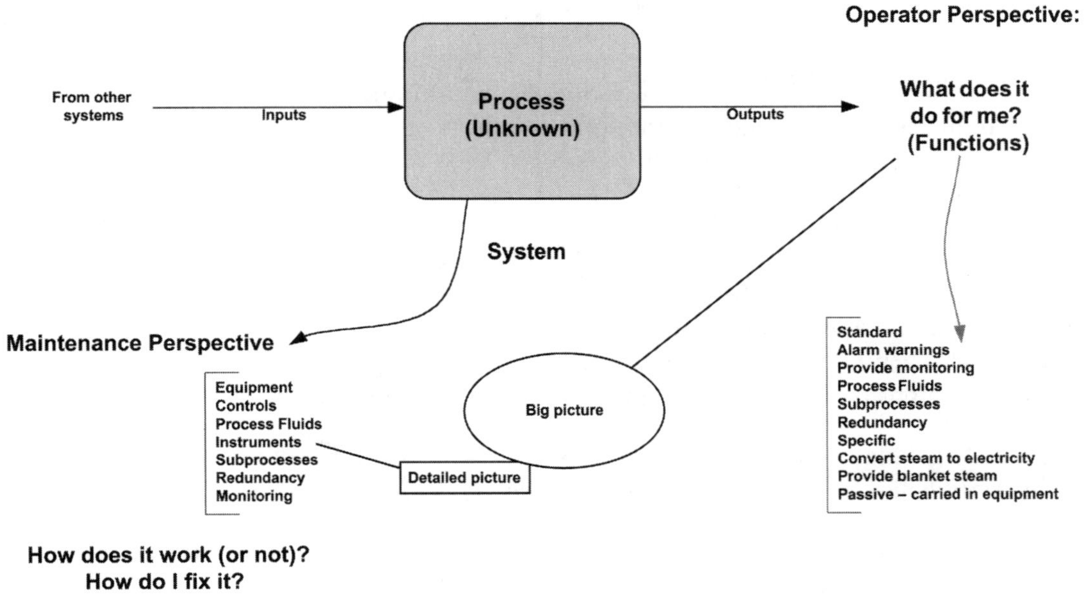

Fig. 1–2 "Black Box" Model

RCM is not a specific maintenance technology or servicing method although it encompasses both. RCM is not what people service or when or how they provide a service although the specific equipment application process identifies what to service and when. RCM does not make the craft or engineers better at their fundamental skills. RCM can suggest expert diagnostic tools for uninformed troubleshooters—where to start.

RCM provides an expert system support that identifies different ways that equipment can fail and the symptoms of impending failure. RCM encompasses traditional preventive maintenance (PM), predictive maintenance (PdM), and corrective maintenance (CM) approaches.

RCM is a risk-management process. It identifies plant-installed equipment risk, supported system functionality, and dominant component failure modes. With insights, operators effectively control risk. RCM selects critical equipment and failure modes so resources are focused. The outcome is a program in which equipment never fails in ways that could have been foreseen and prevented. Resources are used cost-effectively for maximum effect.

Why *Do* RCM?

Reliability is the focus of RCM. *Reliability* is a modern concept; *maintenance* is as old as industrialization. Maintenance craft workers know maintenance intuitively and, therein, lies part of the problem.

RCM is focused maintenance. Over the years, maintenance performers have worked with vague, inexact guidance, little direction except their experience, and with a liberal charge to "do well" and restore equipment. A bankroll allows anyone to perform maintenance; industrial maintenance, however, is a sophisticated dance, best performed when planned. The challenge is to choreograph maintenance steps, aligning them with plant operations to minimize operating disruptions.

RCM is engineered maintenance. RCM provides the best tools for complex industrial facility operations maintenance (*see* Fig. 1–3).

Craft workers know maintenance performance, but do they know the *right* maintenance? Do they know when to do it? Can they show why certain maintenance is correct? Can they discover when it's wrong? (Inevitably, there are times when it's wrong.) Over time, can they incorporate learning? Do they know when they've reached maintenance limits and what the equipment can reasonably achieve under optimum maintenance? (Knowing this determines when to summon engineers, designers, and other specialists to seek product improvement.) Do they know what they can reasonably expect from maintenance, organizationally, with the resources available? Do they view maintenance democratically, autocratically, as a meritocracy, or as something else? Is maintenance an adjunct to operations? Does maintenance complement operations? Are operators involved in providing the maintenance product?

If RCM-based maintenance brings lingering questions to the forefront, other maintenance programs reveal similar issues:

- Total Productive Maintenance (TPM) focuses upon maintenance performance.
- Total Quality Maintenance (TQM) imbues an aura of religion into wrench use.
- Total Preventive Maintenance looks at PM.

All of these contain RCM elements, but RCM differs in one striking way: RCM is *engineered maintenance.*

RCM provides an engineering, technical, and economic basis for all work that an organization performs. RCM establishes both necessary and sufficient conditions for performing work. The RCM process is objective, measurable, and systematic as it selects and performs effective maintenance tasks. Consequently, RCM appeals to organizations with strong engineering values.

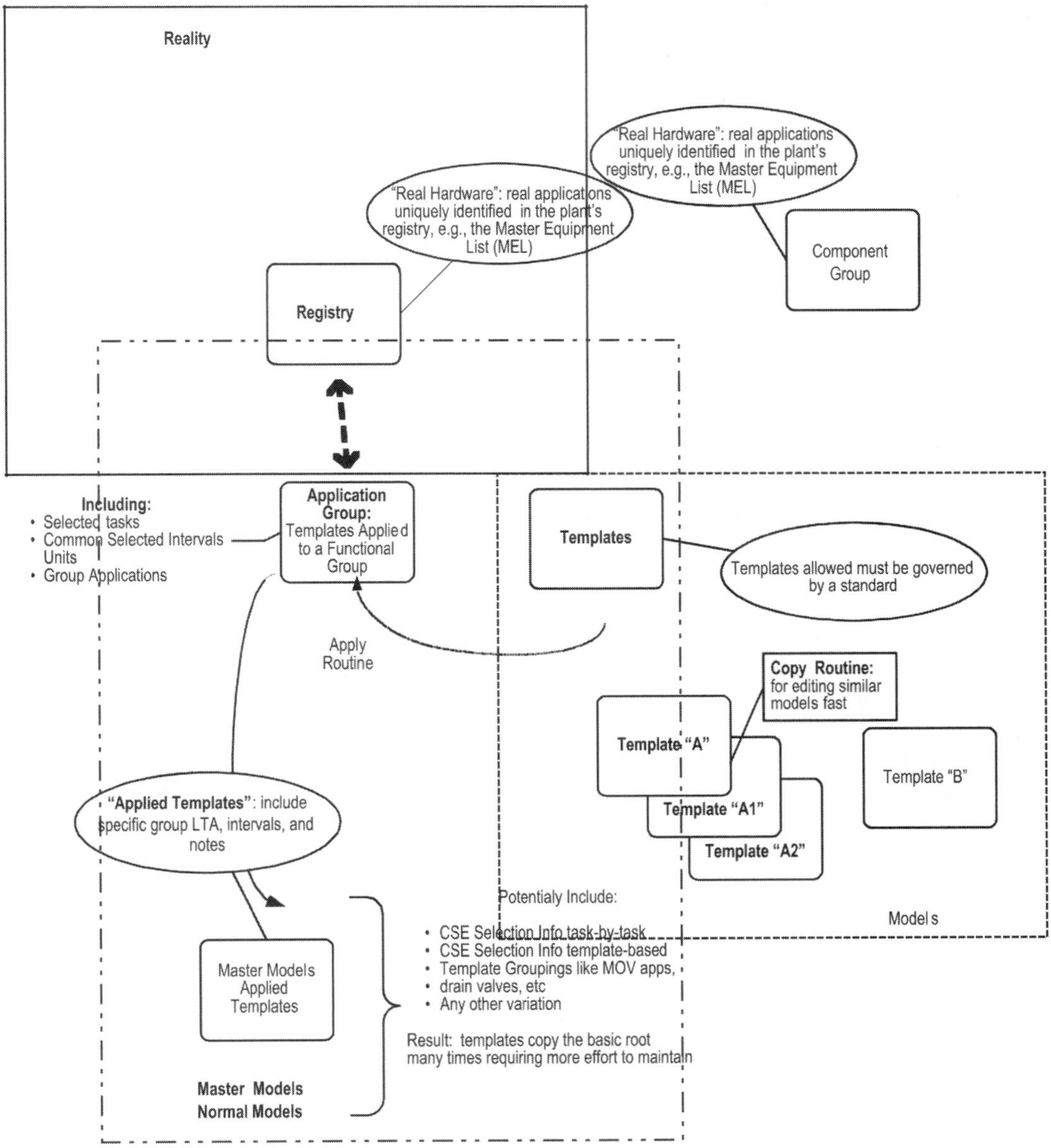

Fig. 1–3 Various Useful Groups in RCM

RCM challenges

The primary concern for industrial organizations adopting RCM is whether they can implement RCM without excessive costs. Pilot studies in various industries show recognition for RCM benefits but concern over resulting analysis cost and its implementation. The primary barrier facing RCM implementation today is cost. (*see* Figs. 1–4a and 1–4b)

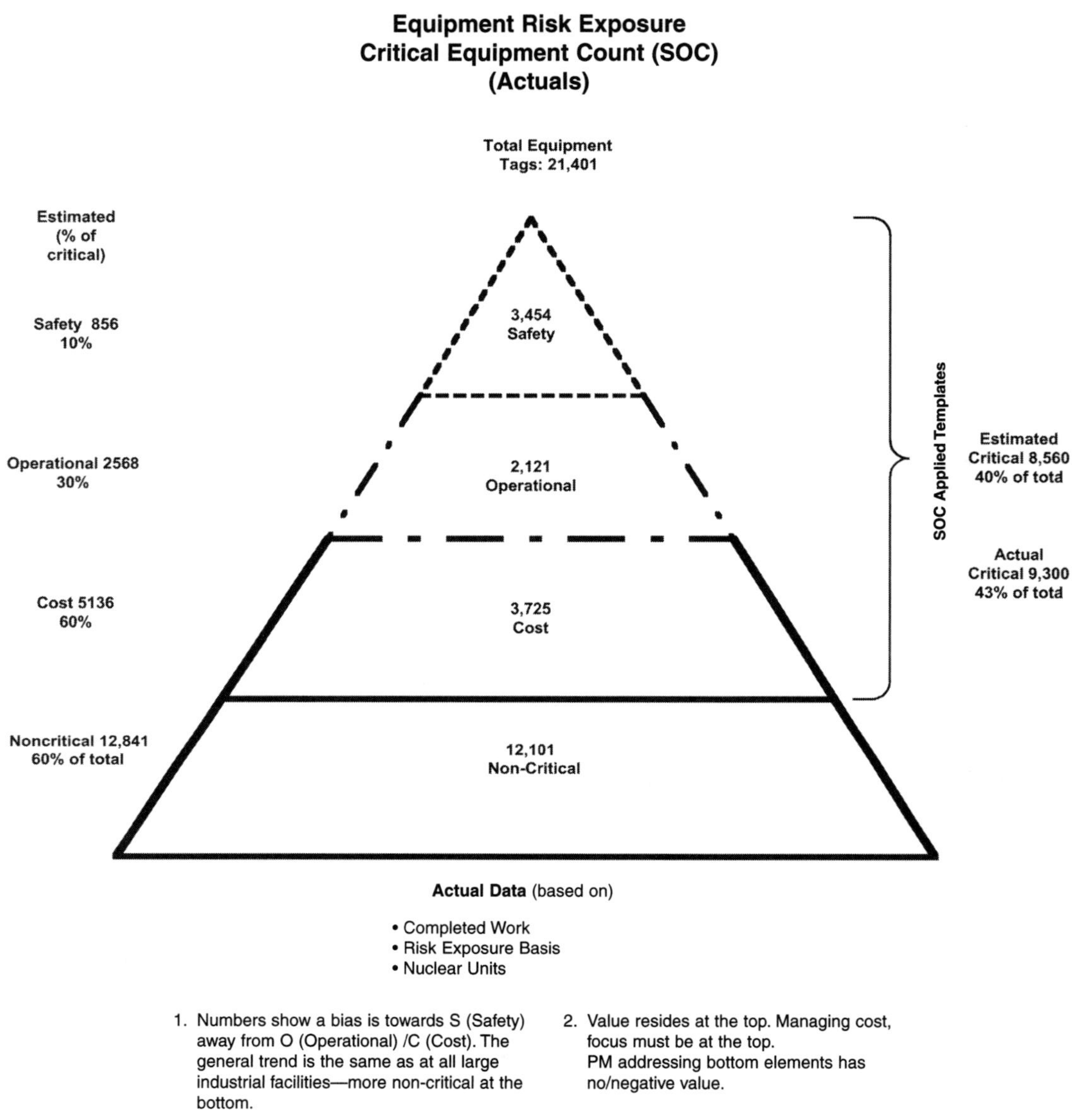

Fig. 1–4a Overview: Equipment Risk Exposure Count

RCM analysis can evolve into engineering studies. Few companies today can pursue blind research, so this challenge is put forth: Can an organization implement useful RCM while controlling cost? What barriers must be crossed to put these reliability concepts into real practice? What are the measurable benefits? What are the real nuggets of RCM? How can RCM nuggets be developed, applied, and used without pain?

RCM involves vision as much as technology. Adopting RCM starts a journey pursuing equipment understanding, learning engineering failures, and learning probability theory, as well as failures, fault trees, distributions, and consequences.

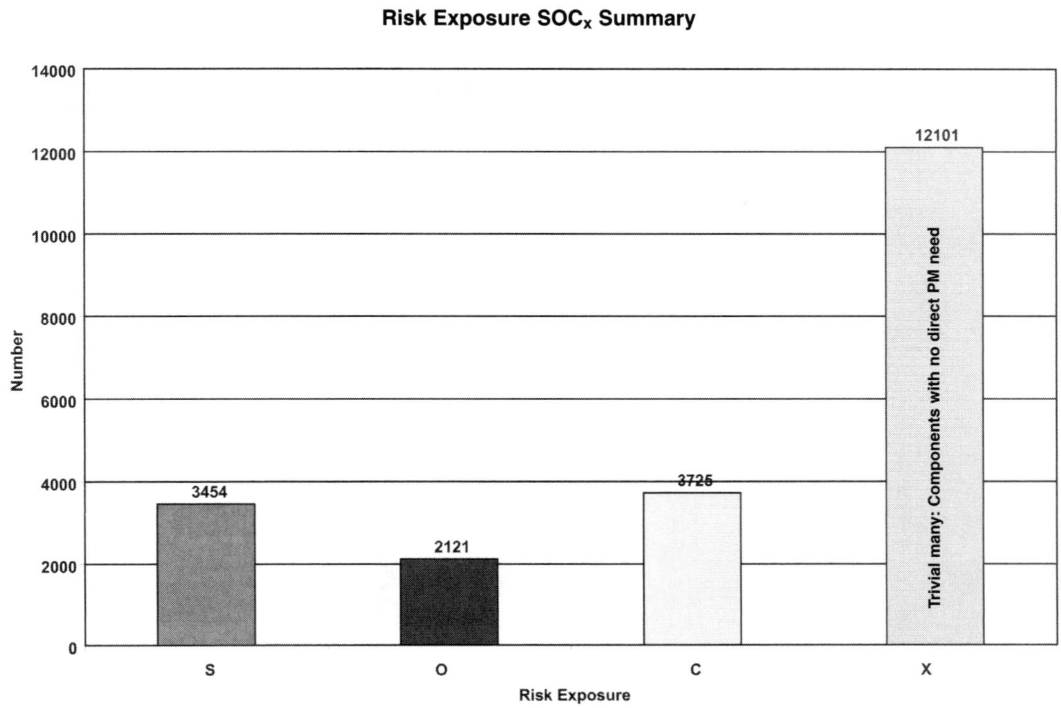

Fig. 1–4b Risk Exposure SOC_X Summary

Failure mathematics doesn't matter as much as concepts. Embracing RCM has consequences similar to embracing total quality management, but RCM is more measurable. Implemented, RCM leads to a new maintenance philosophy that is RCM's engineered maintenance. Companies that embrace RCM must also embrace change, for they will change, shifting work to focus on reliability. Maintenance perspective is the first view to change.

RCM success also hinges on data management. Traditional RCM—like closed-form mathematics—required *crunching*. Other than computers, few can sustain this effort through a large project using handwritten documents or even spreadsheets. No one today would solve finite element analysis with anything less than software. If the answer is the goal, the method that gets an answer fastest, easiest, with the least pain is best. RCM has the same challenge. Users need solutions they can confidently trust, but they need other certainty first. Costs and benefits must be controlled. Uncontrolled analysis flirts with analysis paralysis and can lead to shelf documents for end products. These have no value. Fortunately, modern databases can provide these answers simply. Purists wrapped in traditional RCM's cloak can practice their craft, while those using RCM software databases that interface directly to CMMS get practical results many times over. While the debaters debate, industry moves forward.

RCM is a technology; technology is irrepressible, unpredictable, and irreversible. Industrial applications will move forward using the best available technology.

2 RCM Background

Overview: RCM Phases

Fundamentally, RCM identifies system functions, ranks functions into *criticality importance* categories (safety, operational, and cost [SOC]) based on functions, and develops complete, efficient scheduled maintenance and design improvement plans for implementation by identifying the equipment (components) providing important system functions. Implementing RCM requires system boundary, and functions concepts. Engineers grasp *system* inherently, with underlying utility. Indeed, systems are the fundamental design building block, so RCM begins at the system level (*see* Fig. 2–1).

In its basic form, RCM involves three broad steps:

- selecting components that matter
- selecting appropriate PM tasks for those components
- packaging the results to implement

In summary, RCM

- partitions systems into components
- selects scheduled maintenance tasks
- implements results

These steps are intuitive at a fundamental level, so that almost anyone familiar with a system's equipment, operating risk, maintenance, and cost could draft a basic maintenance program. Learning maintenance—passing through developmental

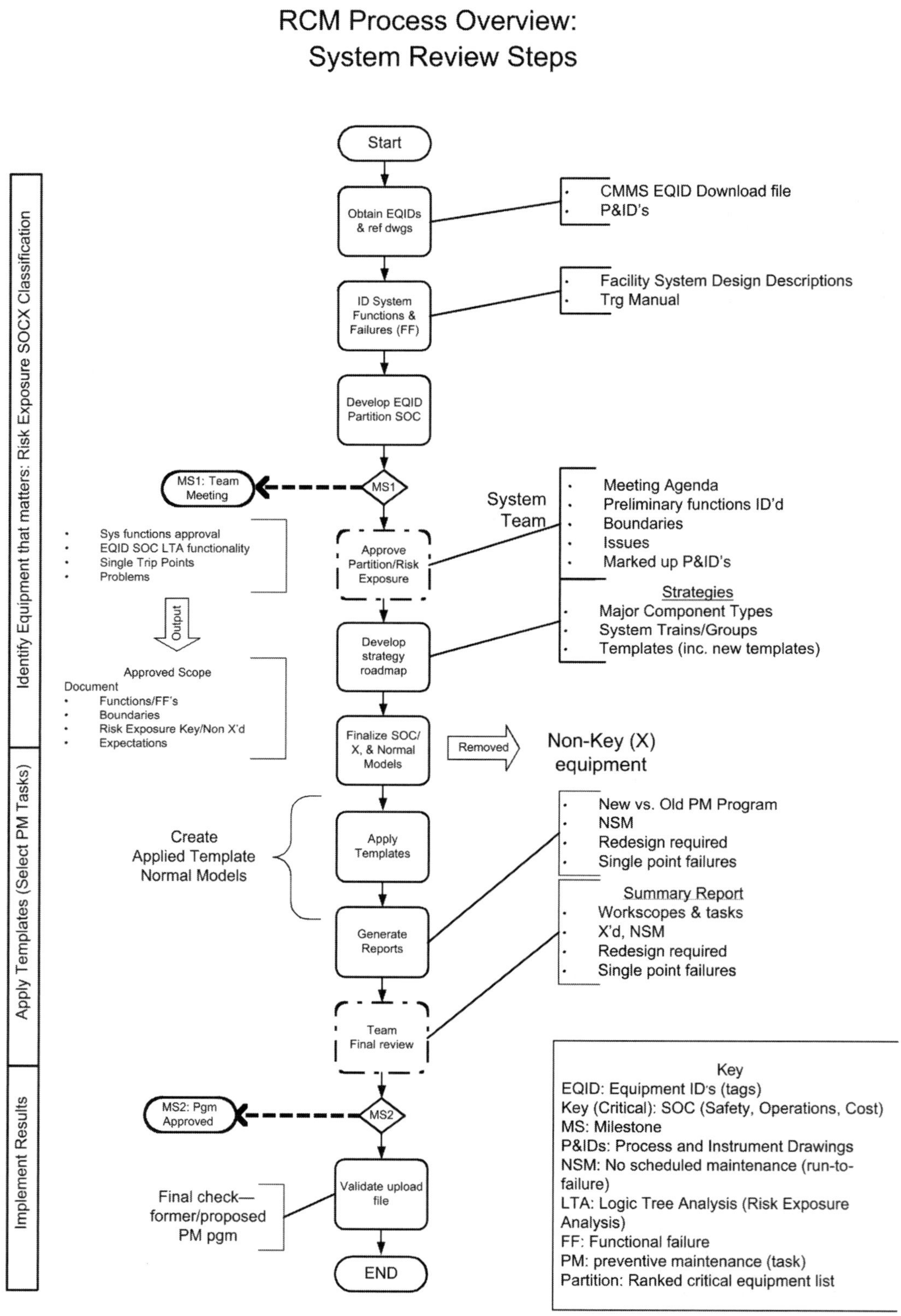

Fig. 2–1 RCM Process Overview

stages—people find they need to learn more about these decisions and their supporting requirements. How can the components that matter be known? Intrinsically, what determines scheduled maintenance task appropriateness? How does that appropriateness evolve over time as technology changes? How can failure-preventing tasks be efficiently performed? Performing RCM reduces to learning how to perform these steps quickly and efficiently.

Components that Matter: the Risk Partition

Critical components are the components that *matter*. They have been called *significant, important, essential,* or *key* in various program formulations. This equipment has the direct potential through single failure to compromise system operating goals (*see* Fig. 2–2). Single-failure qualification is important; equipment that cannot compromise system functions in single failure drops to a lower status. Equipment requiring failure chains to fail in their functions become *non-critical* and are addressed by a program of *no scheduled maintenance* (e.g., run-to-failure). This equipment can be safely and cost-effectively maintained upon discovery of failure symptoms.

Run-to-failure equipment makes up a large inventory of the equipment in any complex industrial facility and supports a strategy of *no scheduled maintenance*. Plants can use run-to-failure equipment to their advantage building a maintenance strategy. Taking advantage of inherent design characteristics reduces maintenance costs and improves reliability. This is the essence of RCM).

Single-failure assumption

Single-failure criteria lead to concise, critical equipment lists. Excluding non-critical equipment from analytical scheduled maintenance consideration focuses effort on the remaining critical equipment. Critical equipment, then, is assigned a risk exposure classification—safety, operations, or cost (SOC). These categories create a system risk profile (e.g., an equipment risk exposure list). This profile ranks relative value, differentiating equipment that benefits from scheduled maintenance from that which does not. Developing this profile is valuable in its own right for evaluating failed equipment during plant operations. This profile also complements the PM plan, providing workers a risk-monitoring guide for prioritizing failing equipment/condition-directed maintenance. The profile also follows the designer's logical thought process, providing margin and spares, extracting plant design depth for maintenance task efficiency. Depth is an asset. Documenting design depth to manage risk exposure is an early step in efficient maintenance plan development.

Effective maintenance programs dictate resource scheduling. By focusing on single failures, programs avoid the complexities and additional resources needed to address multiple-failure chains. In RCM-based maintenance programs, reducing

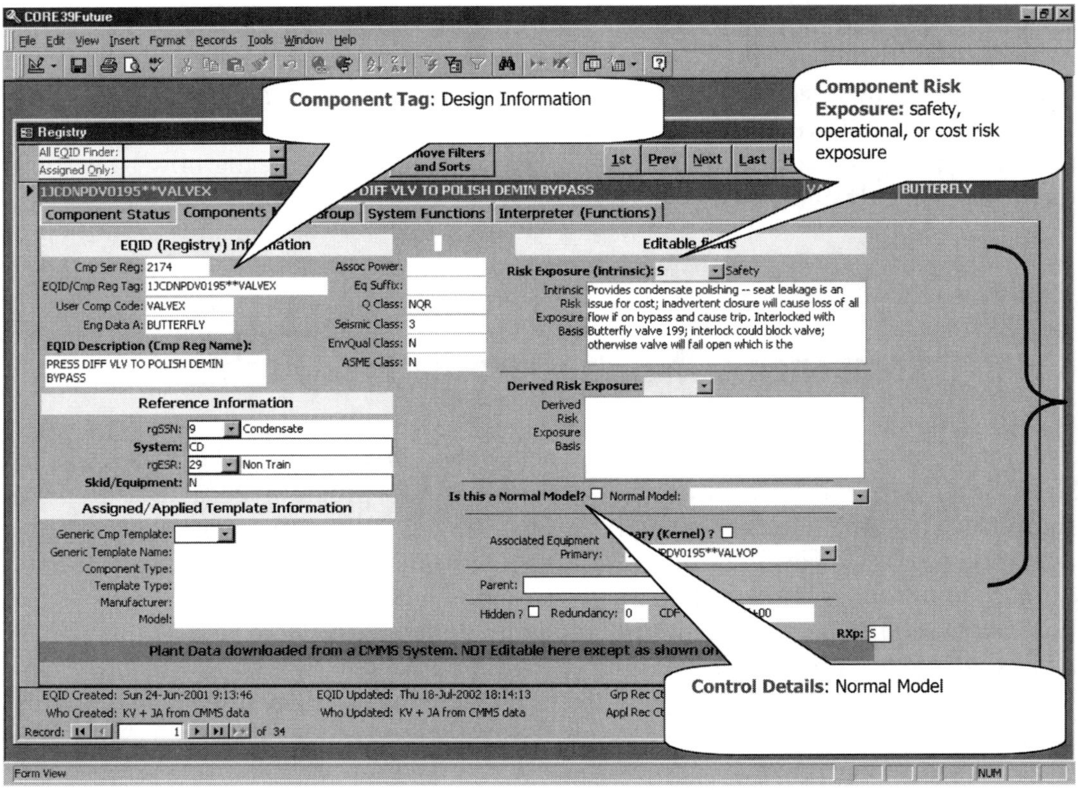

Fig. 2–2 Equipment Registry Entry: Condensate Demineralizer Bypass Valve

scheduled maintenance accompanies an obligation to perform emerging, condition-directed maintenance. Scheduled maintenance discovers maintenance to be provided near-term. When a problem is discovered, maintenance must be performed, or functional failures will eventually result. Plant PM implementation and condition-directed maintenance performance in RCM-based programs must be nearly 100%! Virtually all the scheduled maintenance planned must be performed; virtually all the indicated condition-directed maintenance discovered must be performed within a set time upon discovery. Anything less results in a changed design basis, and multiple failure risk increases.

Maintenance programs can fail from a lack of implementation, resource allocation, or scheduling as much as fundamental strategy weakness. To reap efficient maintenance program benefits, performance follow-through is necessary. In short, RCM is not a fickle-minded maintenance process. Follow-through is mandatory!

Critical classification

Critical equipment identification is based on failure effects. Critical equipment failures that affect safety, operations, or cost-operating goals are ranked mnemonically as *S*, *O*, and *C*. This simple coding scheme reflects three criticality classes (excluding non-critical *X*), which streamline forms and reports. These categories differentiate qualitative failure effects based on three distinct levels and orders of magnitude. Safety losses are at least 10 times as risky as operational ones; operational losses outweigh maintenance costs by another order. These conclusions—cultural values aside—reflect many case studies that reduced failure event consequences to costs (*see* Fig. 2–3).

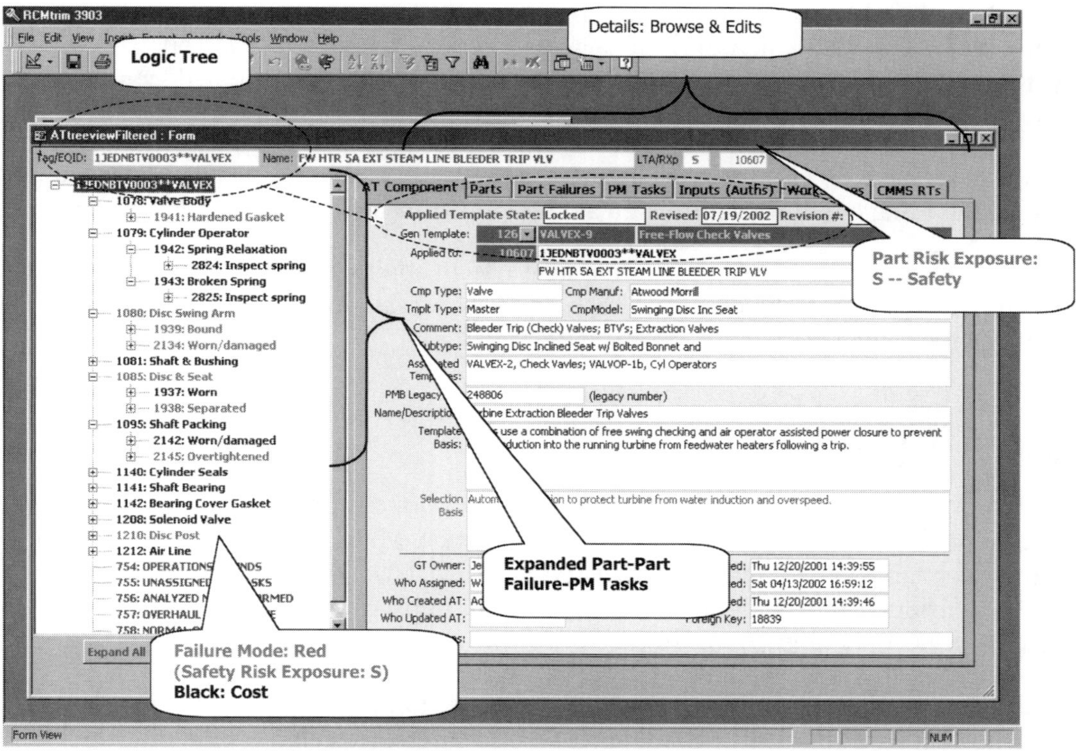

Fig. 2–3 Expanded Applied Template, Tree View

Sub-classification SOC categories depend on failure consequences. Discussing failure consequences in cross-discipline groups develops equipment risk profiles. Discussions capture risks and strategies that improve facility operations and maintenance understanding. Participants reveal and gain fundamental operating risk awareness. Individuals with years of common operating experience still have different failure consequence insights and risk perceptions.

Plant owners, engineers, workers, and operators, should concur on risk exposure assessments. Theoretical operating risks are frozen by plant construction, providing an optimum design baseline. AE design descriptions document system functions that make design failures evident and easy to reconstruct. AE design also reveals intended installed-equipment functionality. Hidden maintenance costs from unfocused tasks compound over facility life. (Hidden maintenance costs are embedded in program assumptions.) Making RCM pay its way requires realizing improved operations. Keeping operational objectives at the forefront assures projects are completed on time with quick payback.

Thumb rules. The analysis process design guidelines and other paradigms result in rules-of-thumb. Manual valves, for example, usually support maintenance. (Operationally critical valves typically have automatic operators.) Manual valve failure consequences are typically cost-based. For maintenance, manual valves can be treated as *run-to-failure*. This rule numerically reduces coded components for analysis review by more than 25%. Therefore, for a typical list of 40,000 installed components, analysis declines by 10,000 items.

All rules have exceptions. One manual valve, for example, may be necessary under special operating conditions to realign an equipment train. That valve could be critical.

RCM thumb rules expand with each new system analysis, case-by-case, superseding existing *ad hoc* PM task selection. Each analysis reveals unique design insights with more reliability benefit.

Methods to develop the risk partition

To develop the equipment risk partition, the developer first needs an equipment list, or *equipment partition* as it is called, for consistency. Many firms know these by their accounting name—the equipment registry. The registry/equipment partition provides the foundation for developing the next logical step—the equipment risk exposure partition or, simply, the risk partition.

Practically, there are two ways to develop the equipment list for risk partitioning. One shortcut is using a partition someone else built. Most large industrial facilities were partitioned at construction for accounting and construction management purposes. Construction requires *takeoff* for work completed to be paid, which requires a partition to estimate "takeoff" completed work from equipment lists. The second, more onerous way is to develop the list. Again, assuming the owner-operator's construction manager had a constructor, who in turn had an AE, the RCM analyst/engineer may be able to borrow and use their work.

Process and instrumentation drawings (P&ID)

New facilities are provided with a process diagram and instrumentation drawings that reflect the design fundamental processes and designer intent. These provide the takeoff points for constructors to procure equipment to construct a facility. They are also the high-level layout of all plant systems and equipment that supports performing RCM.

For RCM, availability of P&ID has another advantage, particularly if they can be copied: They provide visual, highlighted drawings of plant-critical equipment, supporting operations based upon SOC criteria (*see* Fig. 2–4).

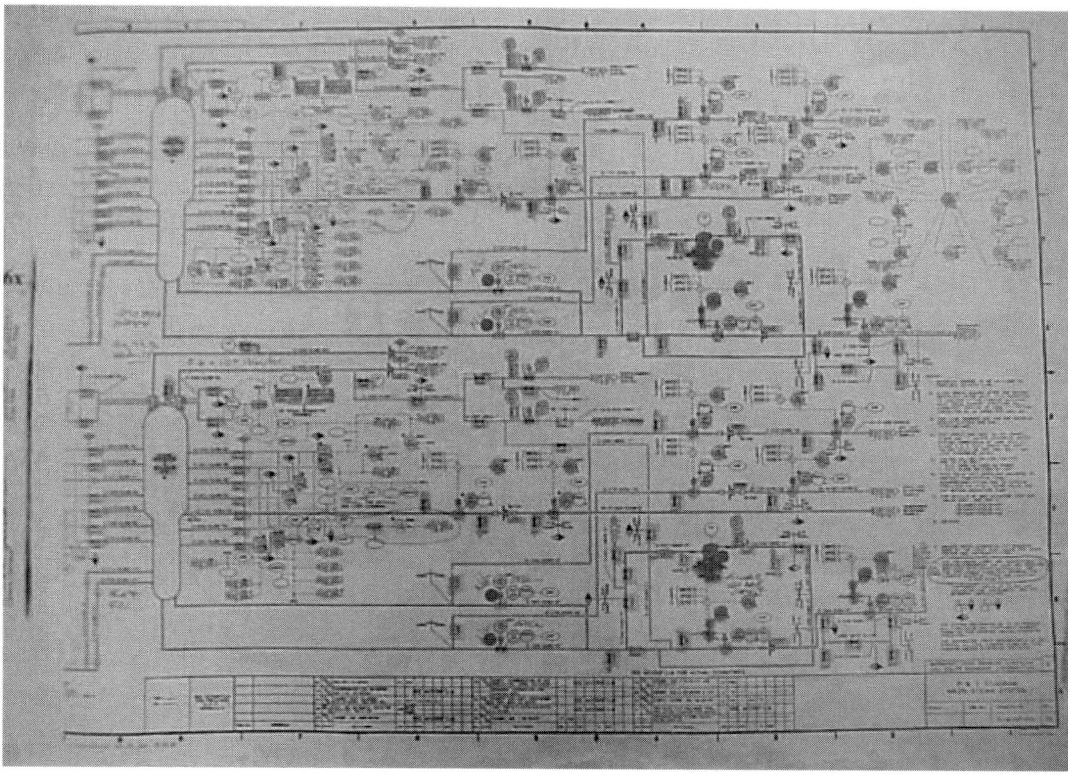

Fig. 2–4 Highlighted P&ID Drawing

The Master Equipment List (MEL) is usually available electronically. Engineering organizations may also maintain a list for design acceptance, construction payment, or regulatory reasons. Nuclear plants must maintain equipment lists for configuration management. Most fossil plants and chemical processing facilities do so for cost and insurance purposes. Where available, equipment lists provide RCM analysis with raw equipment data. Large facilities today use a computerized maintenance management software (CMMS)/equipment asset management system (EAMS). These must have an equipment list for WO assignment. Where these lists are available, they provide equipment identifiers that plant personnel already know.

An electronic equipment list offers many advantages over other forms. Many people are familiar with Microsoft (MS) Excel spreadsheets and MS Access databases or non-relational spreadsheets. As databases, equipment lists can do many things, like retain relational information. That makes using database applications well worth considering as tools for managing and controlling work in large facilities.

PM development databases require an equipment list. Using an existing list reflects the way the plant equipment is already organized. (Few start an RCM PM development project re-numbering plant equipment!) For CMMS/EAMS, using an existing list provides an additional benefit. The existing equipment tag number database key allows the revised work plans to be exported (from the project database) into the CMMS/EAMS upon completion of the project. This seamlessly closes the development cycle.

Appropriate PM Tasks: Template Application

Components determine the next analysis level: identifying dominant failure modes. Vendor operation and maintenance (O&M) manuals suggest dominant failure modes implicitly with recommended PM tasks. Vendors infer failures that could become dominant in their equipment from users' part orders, complaints, and rework or support service sales. They can't actually learn a facility's operations. Vendor-recommended PM, applied indiscriminately, produces lengthy, enumerative PM WO task lists, many of which have little direct applicability, based upon past experiences. At complex plants, common equipment presents low-risk, based upon design redundancy, low utilization, and absence of dominant failures over economic life. Eliminating low-risk equipment from scheduled maintenance consideration early in strategy development (before detailed failure modes and effects analysis FMEA) streamlines analysis minimizing work.

Classifying part failures by failure modes speeds failure analysis and simplifies risk assessment. For example, a pump may fail to start, fail to deliver flow at pressure, or leak. Several part failures can lead to the same pump failure mode. Consider the failure described as *fails to deliver flow at pressure*. For a centrifugal pump, this mode could arise from worn seals, impeller erosion, volute erosion, a single-phased motor, or other causes. A loose bearing guide, however, won't cause this type of failure.

Many part failures are inconsequential. Classifying part failures by component failure modes simplifies system effects analysis performed later. FMEA identifies failure modes that can cause system failures.

Dominant failure modes

Equipment degrades over time by aging as well as random deterioration. *Perfect aging* is predictable; random deterioration requires a probabilistic assessment. Real-world behavior falls between aging and random extremes. Developing effective PM requires the identification of tasks that address dominant equipment failure modes. The reliability engineer must identify component failure modes, parts causing failure modes, part failure mechanism's aging behavior and only then assign applicable tasks. Together, these comprise the elements of the template. (*see* Fig. 2–5)

Applicable tasks must cost-effectively address failure. Failures, to be suitable for PM, must be reasonably likely, addressable by maintenance, and exhibit symptoms that can be found with commercial technology. Theoretical failure modes simply confuse PM performers. In fact, PM addressing failures that don't occur in practice or PM tasks that are ineffective or rely on commercially unavailable technology has no practical utility in industrial maintenance plans.

FMEA identifies failure consequences. Failure consequences determine system effects. System failure effects are more easily discerned, displaying component failure modes with the system's functional requirements. Software displays that present component failure modes with system effects assist in selecting critical failure modes.

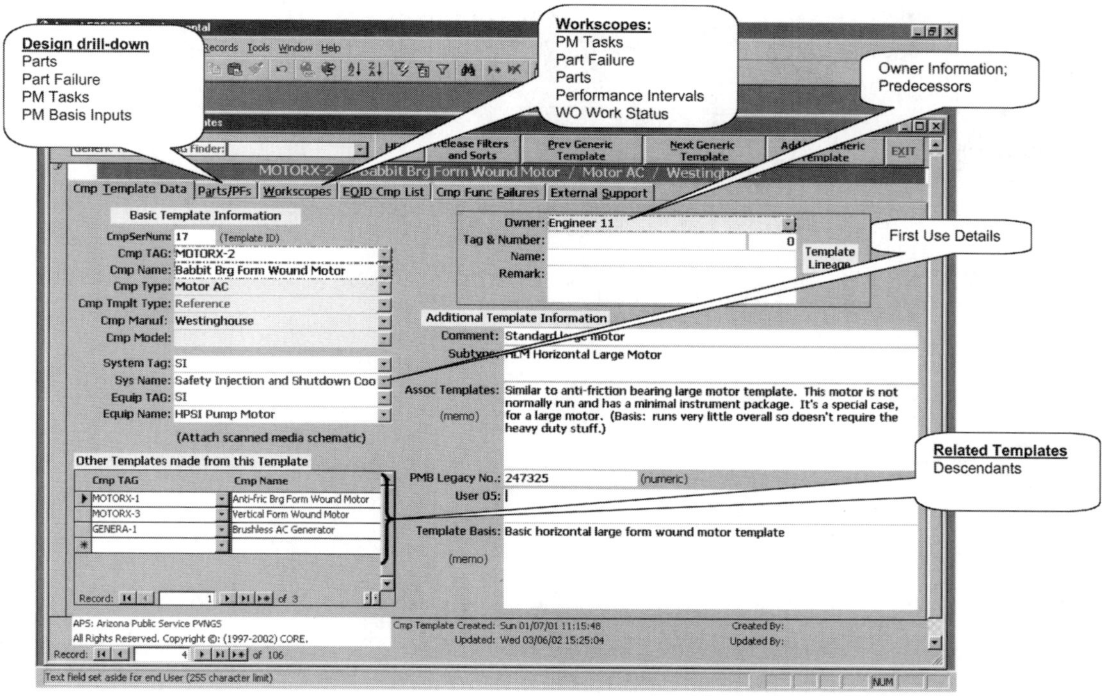

Fig. 2–5 Standard Template Views

PM tasks must address engineering failure mechanisms (e.g., part failure causes). (*see* Fig. 2–6) These are failure processes like stress corrosion cracking, fatigue, or material erosion. Preventing crack failures requires, for example, reworking cracks when a crack failure mechanism like fatigue is present. Eliminating a failure mode is usually beyond the scope of maintenance. Programs need not identify root causes (although that may help), nor do they need to be perfect. Often, failure modes can be managed without resolving root cause. Eliminating equipment failure modes through redesign with root cause analysis is ideal but cost-prohibitive for most non-risky cost or operationally-based failures. Only very expensive failures, or those based upon safety, warrant redesign.

Failure mode development takes time. In studying many equipment types and their failures over years of plant support, the author has concluded that most industrial failure mechanisms are well known. The operational challenge is identifying common failure modes and selecting *applicable* failure mechanism PM tasks quickly on installed plant equipment. Standard templates help to address common components and their parts, based upon known failure mechanisms. Selecting failure modes, critical parts, failure causes, and PM tasks in *pick list* format (based on underlying engineering fundamentals) simplifies analysis. Automating rote analysis makes large-plant, RCM-based PM development feasible.

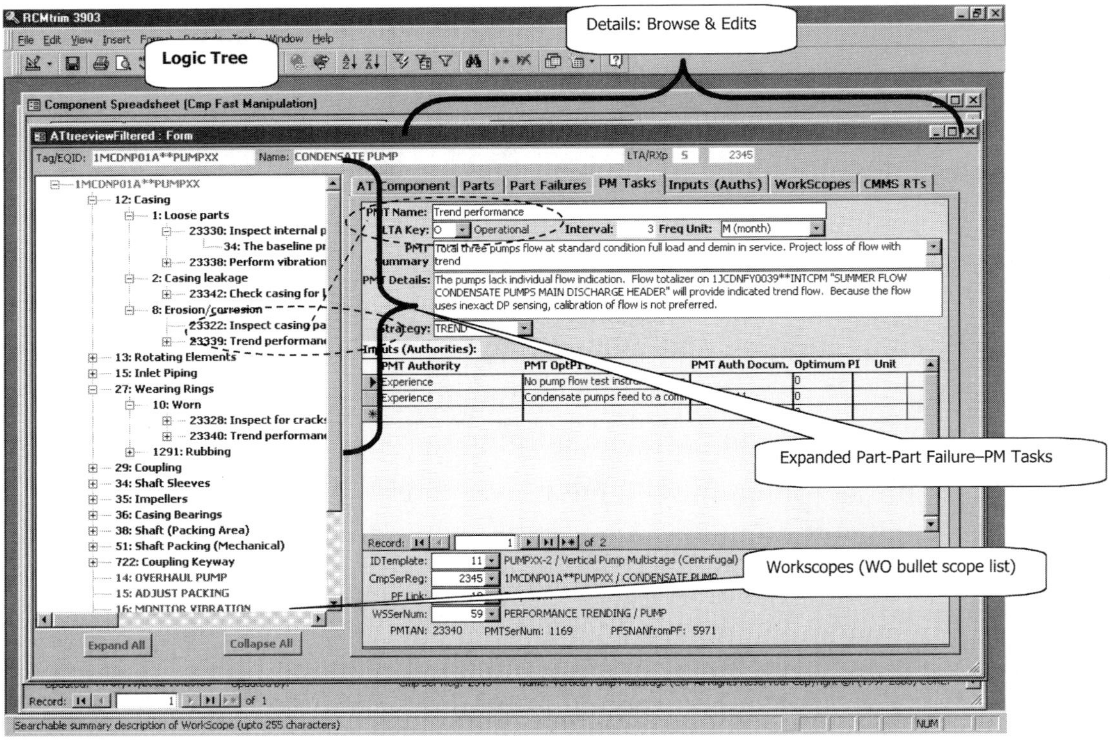

Fig. 2–6 PM Tasks "Tree View" Condensate Pump

Failure management

Once a failure mechanism is known, selecting a PM technology is easy—if one exists. Wall thinning warrants ultrasonic non-destructive evaluation (NDE) wall thickness measurement; pitting can be identified with eddy current testing. Standardized part-focused PM tasks for common dominant failures should be provided. Efficient template development identifies dominant failure modes. Providing generic template models simplifies applied template development, reducing it to choosing parts, failures, and preventive tasks from a pick list of options (*see* Fig. 2–7).

Failure symptoms determine on-line condition assessment options. Instrumentation must identify incipient failure. Selecting instrument diagnostics requires equipment and instrumentation experts. Suppliers provide fault-identifying instruments and alarms to discriminate from *status only* instruments. (The latter are run-to-failure.) Critical fault-identifying instruments reveal hidden failures, and their risk exposure classification corresponds to the risk associated with the failure they reveal. As failure modes, causes, and operating-symptom understanding grows, designers apply more instrumentation and inspection access points to equipment design. Industries that have grown RCM-compliant involve designers at early stages to resolve equipment problems. The resulting design improvements simplify maintenance and diagnostics.

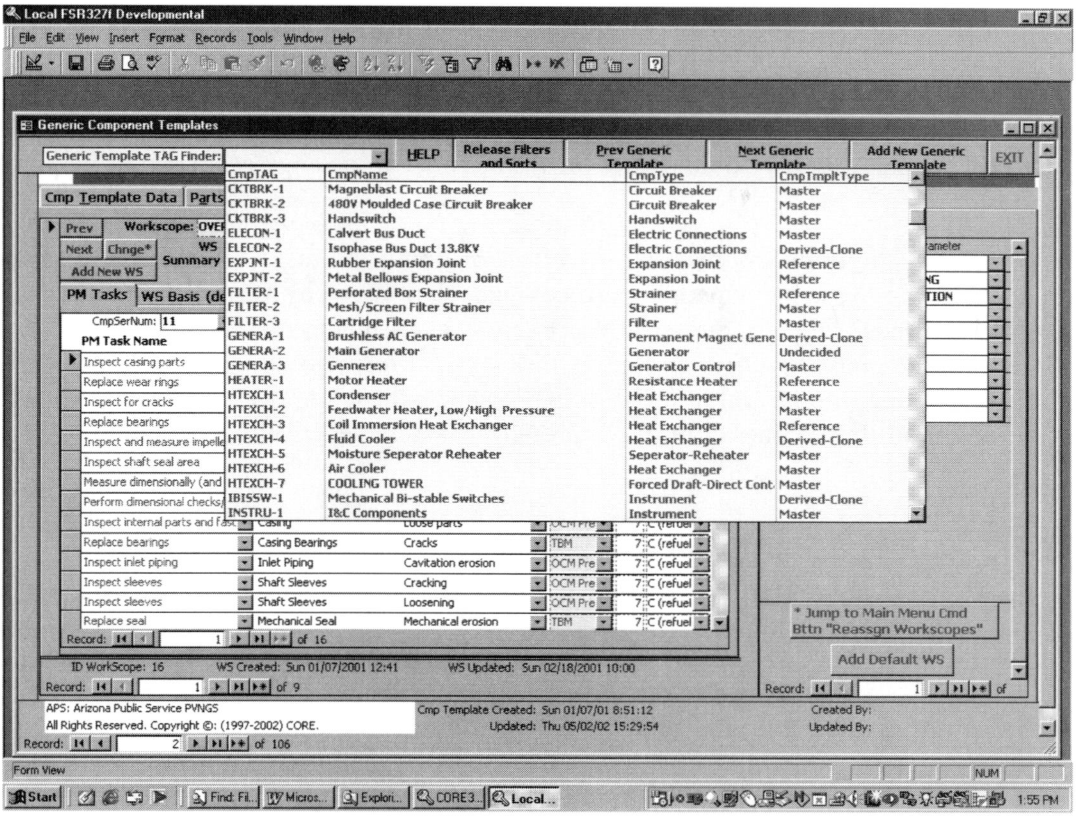

Fig. 2–7 Pick List Template Selection

In contrast, RCM application in generation is still relatively new. Design features supporting RCM-based maintenance are new. Techniques like generator partial-discharge monitoring may be added as retrofits to existing generators based on value, but the life-cycle costs of these features are lowest when they are incorporated designed into equipment.

Applicable and effective

Applicable and *effective* are traditional terms Nowlan and Heap used to describe practical requirements for any scheduled maintenance tasks. They must be technically appropriate; they must work. Additionally, they must also be the cost-effective. *Applicable* and *effective* paraphrased means *technically* applicable and *cost* effective to perform.

The typical challenge for engineering organizations is to assure that selected tasks are technically appropriate. Sensitive, complex sensing-alarm circuits have to work in an industrial environment. This is a tall order, requiring that engineers do the homework when a vendor comes around selling technology. More than a few of these devices spent their lives predominantly on the shelf or out-of-service, once the users learned how *well* they worked. The more practical order is whether one can tote the gadget around the whole day in the summer in 95°F saturated air inside a boiler house. Will it overheat under these conditions? Is it sensitive to coal dust? Will it survive the mandatory initiation dip in the building sump? Practically, a fair amount of equipment goes commercial that isn't quite ready for prime time. Is the engineer ready to help the vendor develop it?

Effective simply means that the net effect of doing the task is beneficial; it beats doing nothing. Most plant people hate cost calculations. Why would they bear such stressful burdens? Few do, and cost effectiveness assessment is largely a leap of faith. *Effective* means *cost effective*, and many tasks would formally drop here if put to this test. Developing thumb rules for cost-effectiveness benchmarks provides a middle ground without burdening all work with onerous assessment.

MSG-3: Task selection simplified

Airline Transport Association Standard MSG-3 (Version 2) provides a process that combines failure selection by importance with scheduled maintenance task-selection logic. Though it provides many complex pages of flowcharts in its original form, the basic selection process is simple, repetitive, and can be summarized on one page (*see* Fig. 2–8). For any item,

- Select the component and determine the failure
- Determine redundancy (if any)
- Determine the risk category
- Use risk the category to seek appropriate PM tasks for the failure

MSG-3 is the fundamental RCM process rendition that has stood the test of time and that remains in force for aircraft maintenance program development. The main points are elegantly simple. The process has many nuances, but the fundamental flow is simple. MSG-3 establishes

- failure control at the component failure mode level
- task classification selection based on a three-tier risk scheme (SOC)
- direct safety (and other) failure limitations
- default actions where safety and absence of *effective* and *applicable* tasks occur

On the finer points of hidden failure, MSG-3 has the time-honed definition of safety hidden failure used as a final test: Does the combination of a hidden failure and one additional failure of a system-related or backup function have an adverse effect on operating safety? By heritage, MSG-3 remains the central, primary standard for

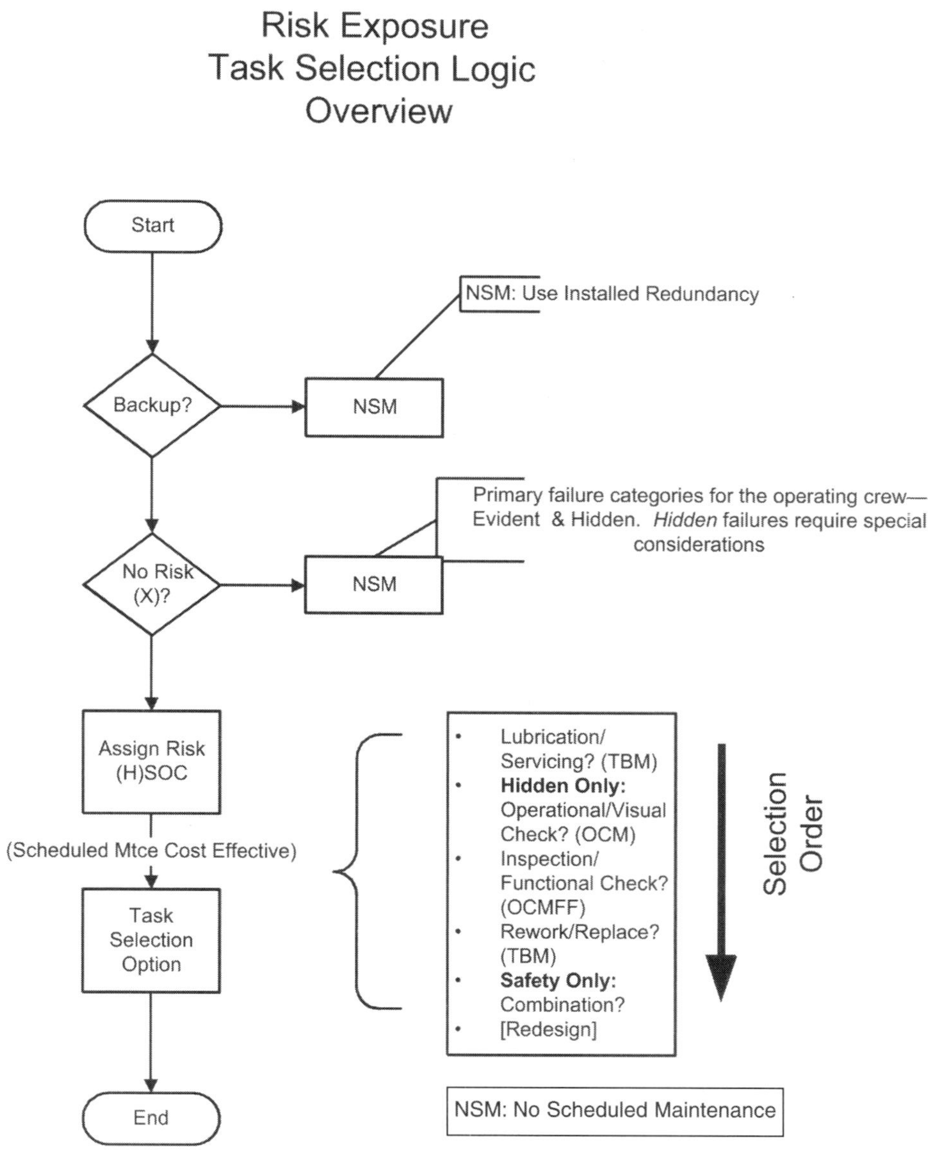

Fig. 2–8 MSG-3 Process Summary

RCM analysis in force today. Its airline industry focus doesn't make its processes invalid in other fields—quite the opposite! It provides a central benchmark to compare with all processes.

MSG-3 establishes a safety criterion that is an important consideration for all RCM-focused programs. A common, RCM deviation in many otherwise rigorous programs is the inability to restrict safety classifications to direct safety failures. This has profound influence on the classification of events as safety, operational, and even cost, as redundancy layers are added. The general thumb rule—that additional redundancy levels reduce failure risk one level and improve operating efficiencies—is not consistently followed.

Packaging to Implement: Upload File Preparation

CMMS/EAMS residence

The CMMS/EAMS holds the scheduled maintenance program. Scheduler-planners manage the CMMS system with a core team of senior, knowledgeable maintenance specialists. Although highly qualified, and the most knowledgeable and experienced craft at a facility, they are not reliability engineers. Many struggle with failure analysis or cost/benefit accounting skills. With the aid of data input clerks, these people manage station work management system information, including scheduled maintenance WOs. Craft workers preplan WOs with target equipment, workscopes, tagout boundaries, risks, crafts, parts, and support requirements identified. Duration and workscope time estimates are provided. *As found/as left* WO entry fields are provided for work documentation.

Difficult step. Uploading final PM optimization results from any improvement effort is tedious, hard work. This may be where a project ends. But this time-consuming, draining effort loads analytical results where they add value—into the CMMS/EAMS. Those tasked to implement RCM optimization results may not own the process and may not even monitor their work. For these reasons, upload steps always present a project completion risk.

Assuring that optimization results get archived provides basic process credibility that any project must achieve. This milestone—and upload implementation barriers considerations and their management of the related barriers—should be considered before a PM optimization project of any kind starts.

Simplification methods. Workscopes assemble PM tasks for concurrent performance. Organizing tasks into workscopes eases implementation. WO tasks consolidated into workscopes present fewer station WO system demands. Many plants assign senior personnel familiar with work practices to assign or *block* PM tasks into organized, easily implemented workscope packages (*see* Fig. 2–9).

Software can simplify task grouping into workscopes. Subroutines must efficiently block and re-block workscopes on demand to support PM development. For an automobile, this would be like shifting check brake pads from 12,000-mile to 24,000-mile intervals. As craft worker and workgroup participation increases, the need to reorganize work iteratively increases. *Point and click* workscope task reassignment techniques automate workscope organization and editing.

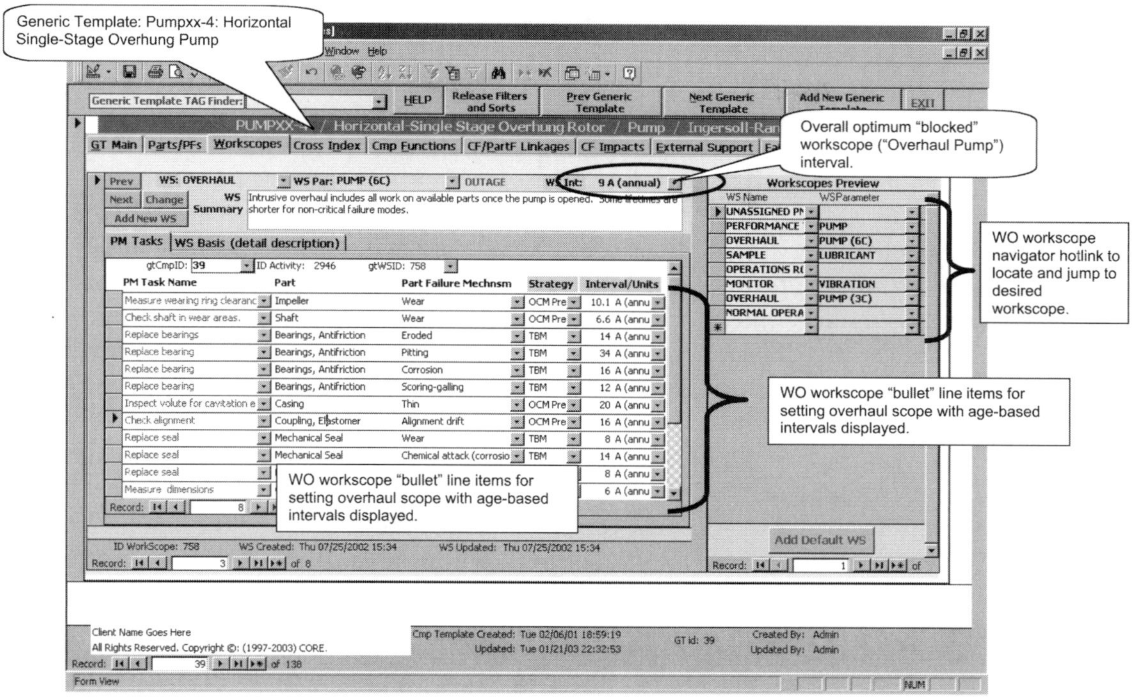

Fig. 2–9 Workscope Tasks Grouped

RCM Steps

Systems

Systems are defined functionally. Systems hinge on functions, so functions influence how engineers look at systems.

Imagine how a designer conceptualizes, plans, and constructs a complex facility. The concept begins with what the expected product is, how it will be produced within the physical, cultural, financial, and temporal project constraints, and what costs and income will be expected. Product output is always the overall facility purpose.

Several hundred years ago, someone wanting to mine, refine, or do practically anything just started doing so. It's not so simple today, and this is why today's facilities have many systems, functions, and requirements. People, the environment, and investors just cannot be used the way they could be in the good old days! Everyone wants assurance that manufacturing and production processes don't create risks or hazards that the public at large will either pay for indirectly, incur implicitly—perhaps through ignorance—or both.

Familiar system and equipment functions are easily taken for granted. Most engineers know systems intuitively and don't want to be bogged down identifying system functions for analysis sake. Engineers want substance. Folks who developed scheduled maintenance programs in use now had these mindsets. This explains why plants generally do more maintenance than they should. Plant engineers are doers, not thinkers.

Systems identify required functions, which reveal their original designers' intentions. Plan layout started with functions; equipment was added as plans developed. A design engineer integrated functional design documents into rough process P&IDs. As conceptual designs were approved, drawings froze. On the drawings, engineers performed ever-more detailed design and provided and labeled equipment. The drawings were used to procure the equipment that would later be entered into a CMMS/EAMS computer system as equipment tags. That's how systems, boundaries, drawings, and equipment lists evolved. Everyone can find these elements from documentation. System functions, on the other hand, are not explicitly known. They are found in the original bid proposal's design requirement section. They represent how the AE planned to provide the prospective owner's desired plant capability when the owner sought a plant design proposal to build the plant.

Functions

Functions constitute the key requirements that define plant systems. Function is lost when supporting equipment fails. Design redundancy determines how much any equipment failure affects functionality and system performance. Functions summarize other document requirements, particularly engineering design requirements, design descriptions, and other high-level engineering documents. For example, functions for a condensate system could be to

- provide boiler startup water at low pressure (<1200 psig)
- provide at least 500 gpm direct makeup flow up to 1500 psig boiler pressure to initiate steam via startup boiler feed pump
- provide 1.5 million gallons per hour normal flow with any two (of three available) condensate pumps
- maintain water chemistry specifications under all operating loads
- provide condenser vacuum alerts at 22 per square inch vacuum (psiv), 16 psiv, and 12 psiv vacuum
- provide hot well normal and alarm level status
- allow condensate dump and drag from hot well to condensate storage and visa versa
- allow online maintenance of an isolated condensate pump

Special functions like chemistry requirements need further specification. Chemistry requirements could be further broken down by O_2 scavenger concentration, conductivity, sodium, pH, and other critical limits that, based on analysis, correlate to metal degradation. Once established, a specification becomes an operating goal to follow (*see* Fig. 2–10). Chemistry functions are largely economic; they allow structures to reach design plant life.

System functions may be identified from unconventional sources, like training materials. Functions that began as system design requirements, providing the system equipment selection basis, may have been augmented as new regulations required new plant functionality. Documents, like operating procedures, user guides, and event descriptions, further supplement design requirements. These may include plant design basis, system function requirements, system functional block diagrams, and lesser supporting implementing drawing, specifications, and design documents.

Together, AE design documents clarify esoteric, often proprietary design details conveying mundane, implicit design lessons gleaned from actual plant operating experience. Experience determines what equipment best meets design requirements. Very likely it was that equipment that was specified for the system.

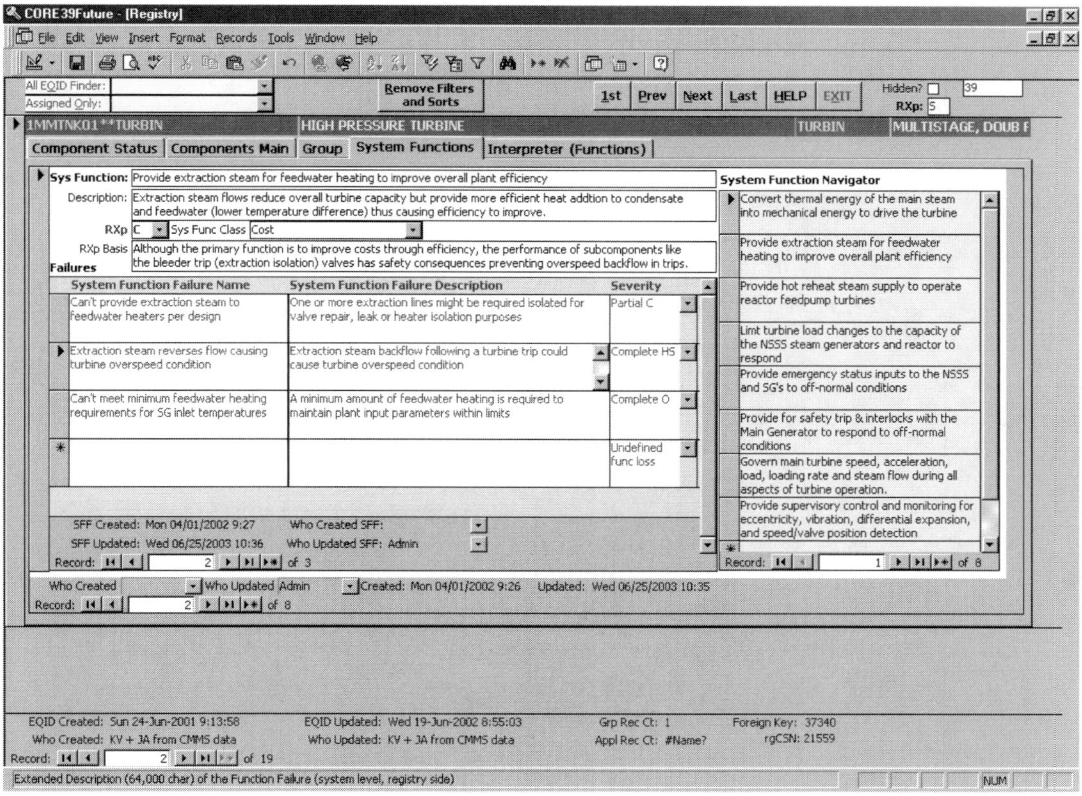

Fig. 2–10 System Functions and Function Failures

Critical equipment

Critical equipment directly impacts system-operating functions. Non-critical does *not* even though it may fail. *Direct qualifier* is important: If the failure of a certain piece of equipment does not directly impact a critical system function, then that equipment drops to *non-critical* status. Non-critical equipment can be maintained as failures occur without scheduled maintenance. Once failing, though, on-condition maintenance must be performed. Once critical function-associated failures, or critical failures, have been identified, analytical focus shifts to the system's critical equipment (*see* Fig. 2–11).

Creating the system's critical equipment risk partition excludes non-critical equipment. Because most equipment is non-critical, this step focuses on scheduled maintenance plan development. In the final analysis, failures are critical. An equipment part failure partition must still be created to identify the part failures of special interest—specifically those that will loose vital system functions.

Hidden failures have special consequences and special consideration. Many equipment failures may be hidden.

Viewing a system from the black box of the control room (CR), failures hidden in the CR may be evident at a local control station. Failure hidden at one level is evident at another; failure hidden in one process is evident in another. Failures hidden while operating a piece of equipment may be evident when the equipment is shut down or

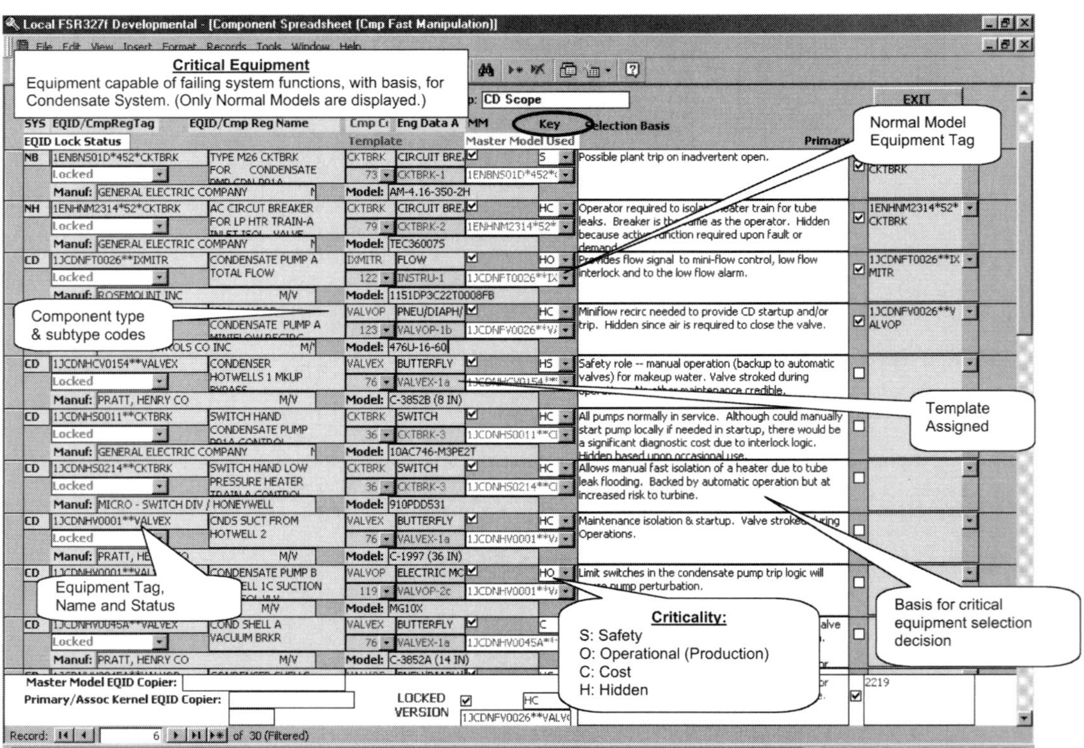

Fig. 2–11 Critical Equipment (Filtered Excluding Non-Critical)

disassembled. Critical equipment selection must be based on the system functionality provided, direct loss consequences, and evidence of loss. Criticality selection must be based on the direct functionality provided—active, passive, or otherwise identified—and on the impact of full or partial loss of that function on the system's function itself. The more actionable, delineated, and specified the system functions, the easier for analysts to evaluate specific supporting equipment.

Critical has many meanings. The meaning intended here is that this equipment has the potential, failing in some way, to directly compromise the system's critical functionality. Focus is on single failures, and how those affect the system's critical functions. Multiple failures are beyond the scope of single failure analysis. Because a well-developed and implemented program removes multiple failure paths, RCM looses no applicability. Design provisions (by codes or license) often remove failures from direct consideration with design strategies such as redundancy or that convert *hidden* critical function losses into *evident* ones. This is the role of a fire alarm for an inaccessible or remote space. For example, the fire the operator might not otherwise see becomes indicated, no longer hidden, for action.

Critical—important, high-ranking, priority, prime, consequential, significant, vital, essential, central, focal, dominant, basic—broadly has been applied to describe many things. Nuclear engineers think of fission chain reactions. Designers think of stress analysis and limiting loads. Pervasive *critical* usage suggests the need for another term to convey the same meaning. Equipment *critical* classification *per se* means very little; *functionality* counts. Most plant equipment fails most often in ways that don't compromise operating goals (e.g., they fail *non-critically*). Therefore, it is possible to remove from scheduled maintenance consideration those equipment associated failures that can't compromise function.

Critical has a direct safety context for airline industry RCM. Nowlan and Heap use *critical* to indicate "S" safety critical risk equipment exposure rank. RCM software vendors have used *significant* and *important* to indicate *critical*. Ultimately, *critical* provides communications clarity: any equipment's intrinsic functionality—the system-required functionality that could be lost by failure—is critical. RCM functional failures are critical. All credible equipment failures that will directly cause intolerable functional losses identify critical equipment.

Technicality

Calling equipment like a boiler feed pump *critical* acknowledges that it can fail the feedwater system and the plant. But a boiler feed pump—a complex skid subsystem with hundreds of subcomponents and thousands of parts—has mainly non-critical, dominant failure modes that won't cause direct pump loss. Therefore, calling the feed pump *critical* alone isn't useful. It could elevate many non-critical failure modes to the same level as the few critical ones. Further *critical* differentiation helps prioritize and manage scheduled maintenance.

Once equipment has been designated non-critical, there is a temptation to not perform scheduled and even condition-directed maintenance solely on the basis of the classification. Not performing indicated maintenance on non-critical equipment—so aptly pointed out by experts—is erroneous RCM understanding practiced by those seeking *silver bullets*. This is not the intent of *no scheduled maintenance*! Not performing any indicated condition-directed maintenance will wreck an RCM-based maintenance program. This choice can cause RCM-based maintenance programs to fail. Sadly, the identification of non-critical equipment for no maintenance (ignoring on-condition based maintenance) is *uninformed* application of RCM that has opposite consequence to that desired. This ironic outcome demonstrates run to failure in the simplest context. There is never any justification for not performing necessary, indicated facility maintenance other than the conscious decision to close a facility. For all these reasons, *no scheduled maintenance* is a better term than *run-to-failure* for non-critical equipment maintenance.

If non-critical redundant functions fail and are not restored, subsequent failure becomes direct. Based on original plant design (and RCM analysis), the non-critical failure that wasn't direct, now is. A conscious decision not to restore failed *non-critical* equipment within the target risk period alters the plant's design basis. The plant scheduled maintenance plan is violated. This is exactly what emerges in old plants that lack a plant design basis maintenance philosophy. These plants suffer an inordinate number of forced outages, and their operating performance records are testaments to failure-based operation. RCM does *not* work in the absence of a maintenance program! Where scheduled and corrective maintenance attainment is not sought, an RCM-based PM program cannot yield results markedly different than any inexact catch-as-catch-can maintenance approach. The ends simply do not meet.

Eliminating non-critical equipment

Differentiating critical and non-critical equipment removes non-critical equipment from dominant failure mode consideration. This has force multiplier effect. Most plant equipment lacks direct failure impact, so identifying non-critical equipment culls that equipment from further consideration. This focuses resources on the remaining equipment—equipment with direct-failure impact that is designated *critical*. RCM's goal is to focus resources where they have greatest benefit. Screening unimportant equipment from scheduled maintenance consideration focuses on the critical few (*see* Fig. 2–12).

Secondary failure

Secondary failures can indirectly cause functional failures. They also cross system boundaries. For example, boiler-corner sootblowers, blowing in a full arc, cut corner wall tubes; tube cuts are secondary failures in the boiler steam system. Sootblowers keep heat transfer rates high, keeping efficiency up, but they must not cut boiler tubes, which must retain steam/water pressure integrity.

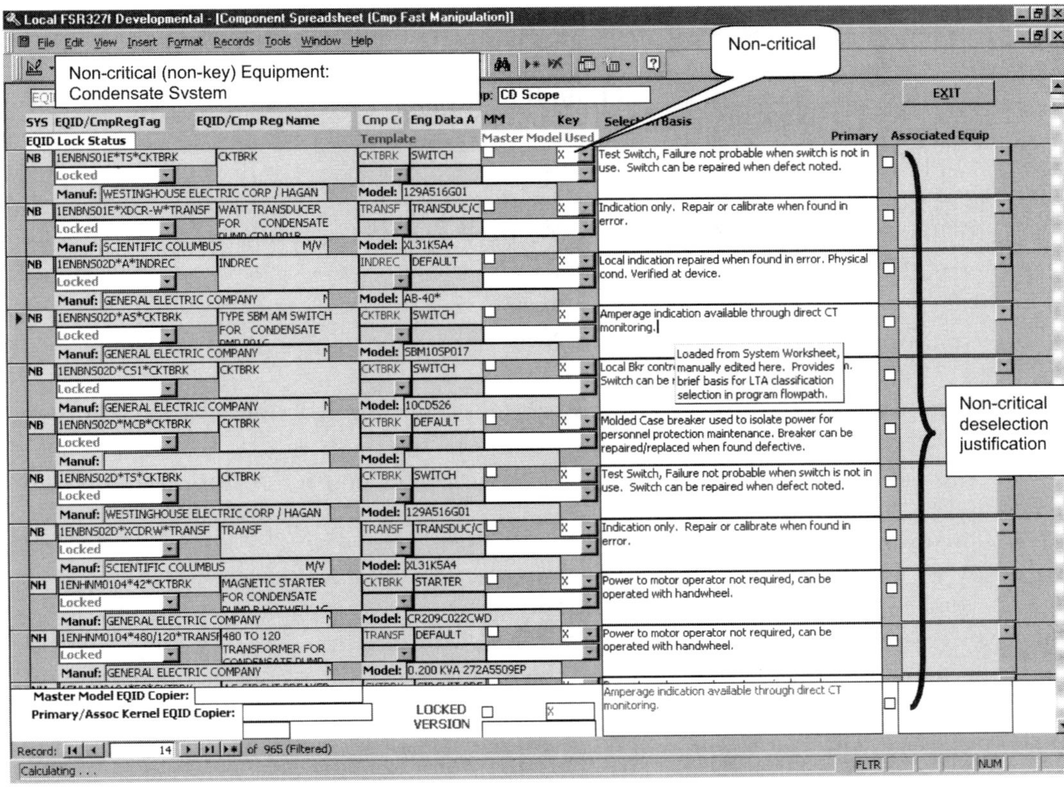

Fig. 2–12 Filtered (Excluded) Non-Critical Equipment and Basis

Preventing boiler tube failure attributable to blower cutting requires managing blowing patterns to prevent patterns that are incorrect or excessive. Blowing patterns contribute to sootblower function. In a unit with hundreds of sootblowers, maintaining boiler heat rate efficiency is the cost-related *should maintain* function. (Sootblowing *should* maintain boiler design temperatures, pressure, and heat rates.)

However, tube-through-wall cutting is another failure mode. Boiler tube leaks introduced by blower tube wall cutting from incorrect blowing patterns are secondary tube failures. From the boiler *contain steam/water* function perspective, tube failure is critical. Loss of boiler tube wall integrity (like any other boiler steam path leak) shuts down the boiler due to tube leak steam loss. Sootblowing system blowing pattern performance is consequently critical to the boiler system for what it must not do: cause indirect boiler tube leaks. This newly identified "O" critical function blowing fails with loss of blowing pattern control. No amount of PM on the secondary failure's parts or systems will ever effectively remove the blower tube cutting cause of the failure introduced by the sootblowing system's blowing pattern functional failure.

Specifying blowing patterns is as critical as blowing for critical blowers. Critical sootblower function requirements at the sootblower system level include specifying blowing pattern to avoid excessive blowing and tube cutting. It's as important that blowers don't blow excessively or out of pattern, as to blow at all.

Infrequent major losses often involve secondary failure. They include

- unexpected problems that should not have occurred
- support systems that were involved
- primary failures that went unrecognized because secondary failure potential wasn't appreciated
- support systems specifications that may have been absent or incorrect
- systems interfaces that were crossed (sootblowing to boiler steam/water)
- function(s) that were unspecified or overlooked in one or both systems
- interactions that weren't appreciated

The sootblowing system should maintain clean tube wall circuits but should not cut boiler sidewall tubing. The practical difficulty is identifying *all* system functions, including passive and *shall not* function requirements that a system should not cause, while developing system functional descriptions. Operating experience plays a big role in uncovering these functions. Experience identifies latent, unappreciated functions once a new system has had a few operating years of service.

Specifying a *should not do* list for a system function and its equipment is as important as specifying what it was directly should provide by design. *Should nots* are easily overlooked, learned by experience, and frequently unanticipated—especially as secondary failures—until they happen. *"Oh, my God!"* events may be unforeseeable until they occur.

Consider main turbines steam extraction systems. They improve cycle efficiency. They also should not destroy the very machines whose efficiency they are trying to improve under any operating condition. Yet, that's exactly what happened on the first turbine equipment outfitted with extraction lines. The first few extraction turbines lacked check values and trip isolation equipment installed to prevent steam reverse flow *on trip*. In the late 1930s, two catastrophic turbine losses from trips occurred, followed by runaway turbine-blade shedding and missile ejection.

Finding all the things systems shouldn't do but might do is an inductive exercise in the absence of plant or industry operating experience. Many things can happen. Some aren't directly preventable secondary-cause events.

Function partition, failure modes, and risk

Critical equipment's failures correspond to tasks that prevent failure or mitigate failure effects. Identifying parts whose failures can cause direct component-function loss focuses on how equipment parts lose functionality, and how part failure consequences affect component and system function. RCM defines failures to find effective and applicable preventive maintenance tasks or to design failures out altogether. Work-process knowledge, technical expertise, and industry-provided diagnostics awareness help the analyst identify applicable and effective scheduled maintenance tasks.

RCM analysts must understand equipment/components as subsystems, then understand their constituent part failures as failure modes and mechanisms. Understanding subsystems means knowing component-level functions and functional failures. Armed with functional partition they can address failure modes. This approach differs radically from traditional PM techniques: traditionally, one surveyed vendor-recommended O&M manual tasks, selecting those that "looked right." When unambiguous or emphatically presented, perhaps assisted with other operations criteria, the analyst selected those that "looked right"—isolated from other contextual factors like service and risk! This seat-of-the-pants approach is much less exact than analytical methods invoking RCM.

Fault trees and root cause analysis

Other processes such as fault tree analysis (FTA) and root cause failure analysis (RCFA) influence RCM failure analysis (*see* Fig. 2–13). FTA and RCFA must be applied consciously. RCFA has wide application and utility in quality assurance. RCFA prevents failure through elimination; when failure reoccurs, RCFA looks for cause(s) in order to eliminate failure. For random failures, redesign may be effective. However, if redesign cost exceeds replacement cost, scheduled maintenance is more cost-effective. Redesign costs typically exceed costs associated with performing maintenance. Where design anticipates maintenance, it's unlikely that the redesign option will improve upon maintenance program strategy. Effective maintenance does not require RCFA. Scheduled maintenance itself acknowledges that failure can be managed. Redesign should only be introduced for intractable design and maintenance problems. MSG-3 logic reveals when redesign is necessary.

FTA *bottom events* initiate a failure chain of events up the fault tree, potentially causing function failure. The *bottom*-initiating event causes a functional failure *top event*. All possible event combinations that yield a top event define a set called a *cut set*. Functional failures develop in many different ways, and many, theoretically, involve multiple failure paths. RCM precludes multiple failure chains that lead to top event functional failures. The reason is profound: FTA is complex; RCM is simple.

RCM controls initiating bottom-events by removing single failures. FTA/RCM relevance is the initial event failure: a part failure (*see* Fig. 2–14). Part failure always initiates every preventable component failure. Part failure may *cause* component failure. The failure mechanism underlying component failure (wear, elastomer deterioration, or metal fatigue, etc.) initiates failure inherent in design. It can only be eliminated by redesign, which, typically, is an expensive proposition (*see* Fig. 2–14).

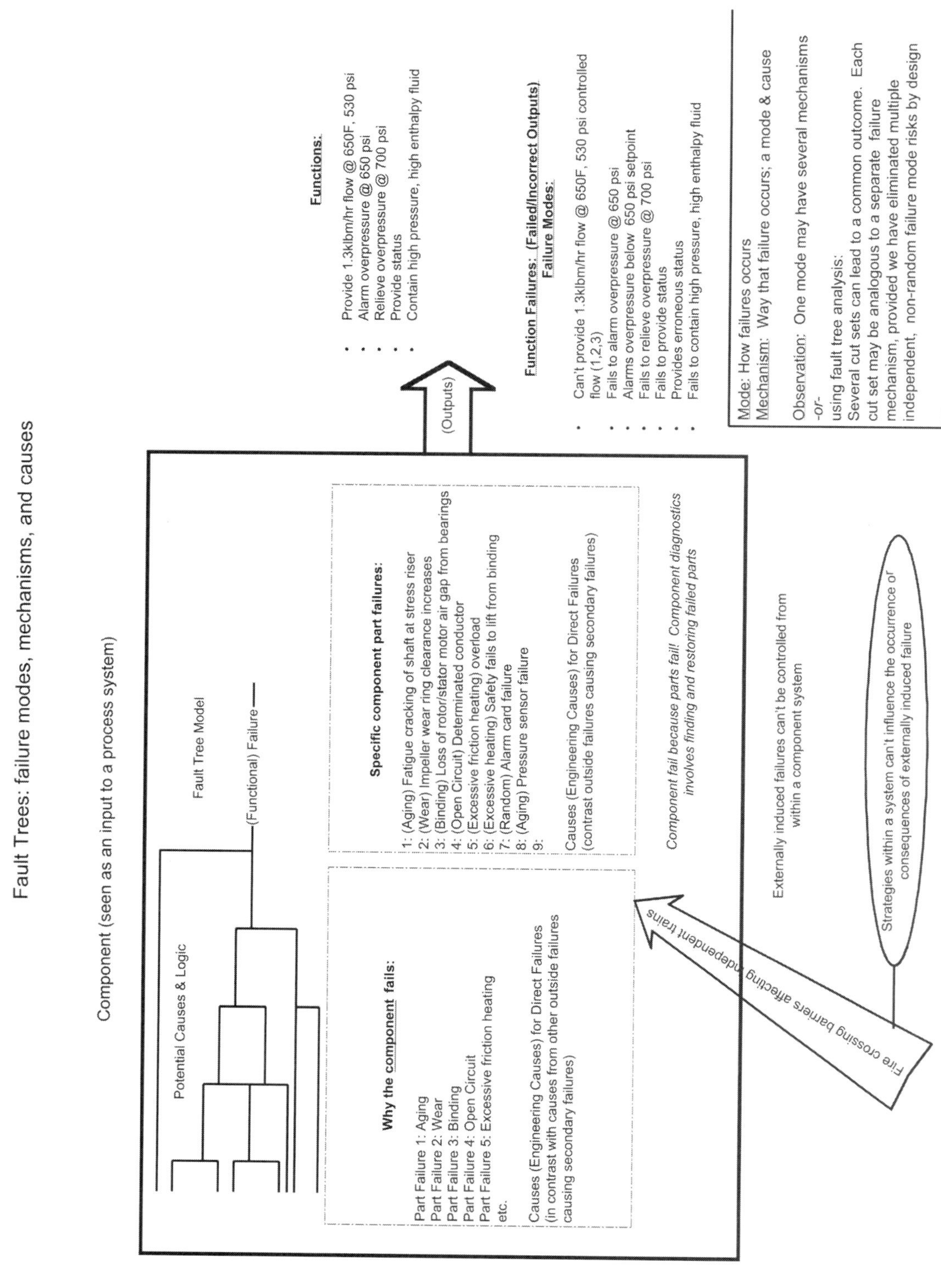

Fig. 2–13 RCM and Fault Tree Analysis: Failure Modes, Mechanisms, and Causes

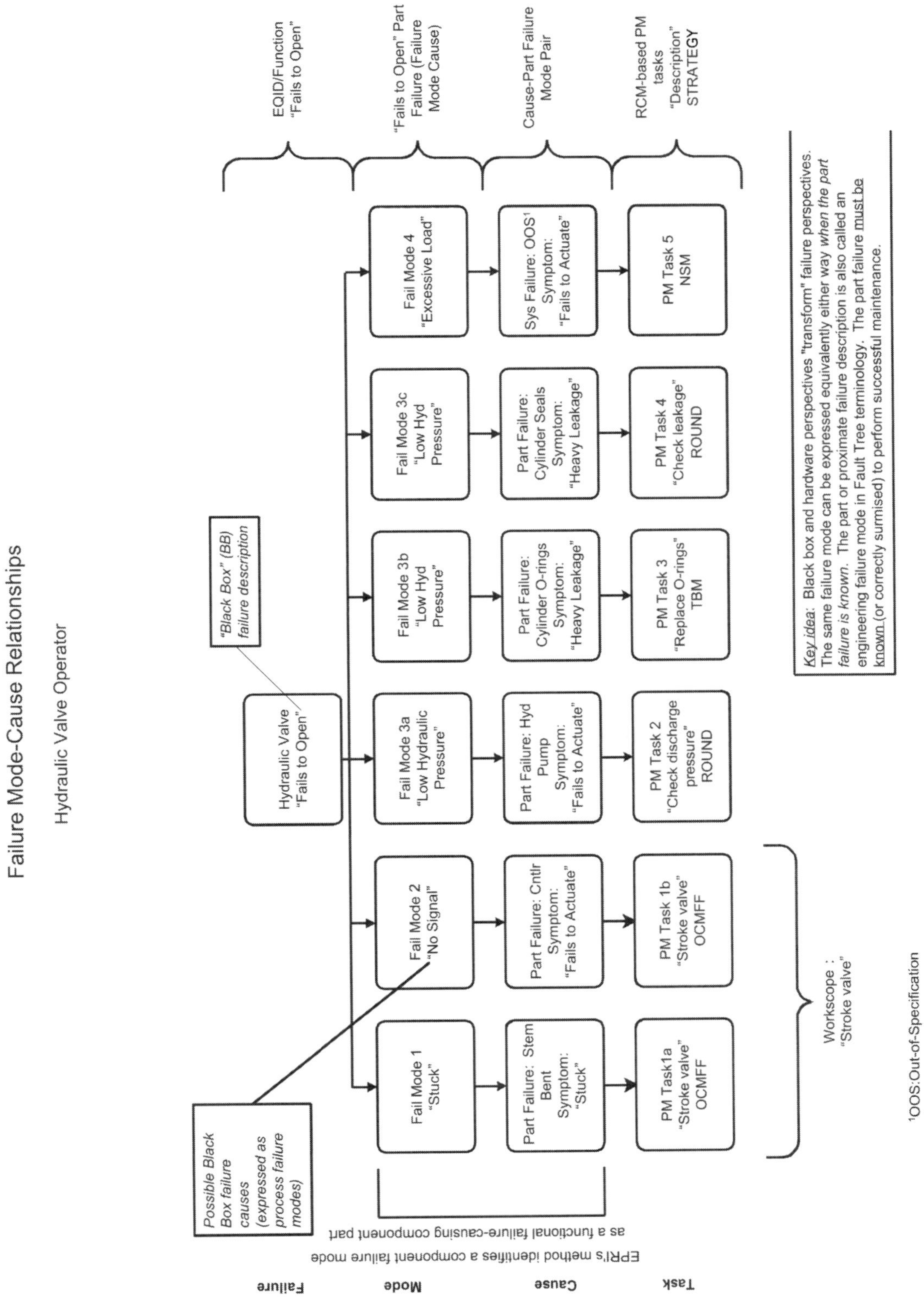

Fig. 2–14 Part Failure Causes

Part failure mechanisms are also called *engineering failures*. This term suggests the inherent physics of the failure and its design implications. In fault trees, an event is an occurrence, anticipated or not, like faulty trip, spurious trip, or trip without demand. Uncertain initiating events are more serious than predictable ones; they are harder to control.

Failure modes describe how failure occurs, not what the cause is. In a failure fault hierarchy, the mode (e.g., effect) at one level is caused by the next lower tier failure effect, which is another mode at the lower level. To minimize confusion, part failure modes are referred to as *mechanisms*.

Part failures cause component failure. Failure modes explain how components fail to provide functions. Examples include

- load failure
- start failure
- low output
- leakage
- indication failure
- throttle failure

Component-level failure modes are brief, succinct, and, for equipment/component templates, need not address performance requirements. Component failure modes broadly express how component function is lost; therefore, they closely tie to functions. Dominant failure modes (DFM) reflect service application and are *reasonably likely* to occur without scheduled maintenance. Analysts find reasonably likely DFM criterion difficult to assess. Experienced teams coached by experts, using statistics gleaned from CMMS/EAMS WO failure history, develop the best PM development DFM statistics.

Physical hardware hierarchy ends at parts. Part failures cause component failure events. A failure mechanism should identify part-aging physics with superposed random effects. Unless fundamental failure physics change or redesign occurs to remove it, a failure mode intrinsically envelopes a physical aging or random deterioration process. For example, lowering environmental stress (temperature) or raising a part's inherent resistance (material change from a Buna-N to Viton elastomer, for example) could favorably influence elastomer aging. Fundamentally changing the physics of Arhenius temperature aging is *not* an option; elastomer aging based on temperature is physical law, inherently constraining, unchanging, and timeless.

PM task selection

PM task selection depends on understanding failures at the part level. MSG-3's task selection logic has become known as *logic tree analysis* in part because it is based on a series of analytical questions in a *tree* format. These analytical question responses dictate the failure risk SOC and prompt PM task selection when appropriately asked in logical sequence. The first question is always, "Is there a task that effectively prevents this failure?"

The task progression selection sequence always considers (in order)

1. light servicing
2. condition monitoring
3. failure finding
4. time based
5. combination (safety only) of Nos. 1–4

The No. 5 selection is significant as a deviation from the established pattern. RCM tasks address failure based upon symptoms with a high degree of certainty. In the case of safety, where one technology or method is ineffective, a task combination is allowed if it reduces the failure occurrence probability to an acceptable level. For other cases—production and cost, for example, those risks can be tolerated since these reduce ultimately to economics; for safety, redesign is required or the failure becomes a showstopper.

The first option is always light servicing—tasks that can typically be performed at the operations level. The last task (safety excepted, as listed previously) is time-based/hard-time maintenance. The progression moves from light maintenance, servicing and condition monitoring toward time-based, overhaul-type activities. Where time-based tasks are necessary, they are simplified, like entire subassembly replacements.

Risk exposure

Components provide functionality based on their system design role and intrinsic functionality. RCM identifies components that can fail directly (because they are restricted to dominant failure modes) based on SOC consequences. Component classification by system function with exposure risk identifies the *critical few* components that receive detailed analysis. This segregates equipment that requires scheduled maintenance from that which does not. The latter may be termed *run to failure*, since it has no scheduled maintenance. Critical cost, operations, or safety consequences determine what to do once dominant failure modes are identified. Complex, logic tree-like diagrams depict this resource allocation logic, so it is appropriately named *logic tree analysis* (LTA). Note the similarity with *fault tree analysis*.

Dominant failure selection requires skill and intuition. It can be subjective. Experience in identifying dominant failures helps, but hard statistics help even more. Dominant failure modes present themselves in failure history. Evaluation of the failure history is never inappropriate; but where the analyst is confident in his understanding of equipment, it may be unnecessary. Equipment groups can validate dominant failure selection. Equipment group selection requires experience and skill; but when done correctly, it provides failure insight and allows analysts to identify unique equipment populations with common failure behavior.

Some failure modes exhibit *potential failure* (PF) intervals and can be recognized using predictive technology to identify emerging failures. PF intervals allow the identification of an emerging failure in enough time to respond, thereby preventing failure. With potential failure modes, physically discernible features precede failure by a quantifiable interval. Where potential failure leads to function failure (FF), preliminary inspection reveals the potential failure. Maintenance specialists can then identify failures with known as PF-F intervals with adequate lead time to avoid functional failure.

Adequately known PF-F interval is the fundamental criterion for applicable predictive, on-condition maintenance task selection. An excellent maintenance program identifies and prevents failures from PF precursors that would otherwise develop into function failures. For example, HVAC switchgear-supporting systems reduce aging stress. Many age-related switchgear failures would develop into dominant failures if supporting HVAC systems weren't maintained.

A significant part of a plant engineer's job involves the development of suitable technology for failure discovery. This requires a combination of coordinating with vendors to find suitable technologies and statistical tracking to locate areas for development by value. It is also helpful to establish a close relationship with the crews who service and operate the equipment in order to discuss clues about symptoms that provide insights for the technologies to select for failure discovery. These activities are the continuous improvement /research side of plant operations.

Equipment failures and failure modes establish systems and equipment relationships. Interdependence causes maintenance program complexities and potential oversights. Equipment programs coexist with a balance of maintenance programs. Eliminating supporting maintenance program infrastructure affects ongoing equipment failure modes and necessary maintenance intervals. Wholesale abandonment of a supporting equipment or system program often has unintended consequences. Secondary effects are difficult to foresee, so they must be avoided outright—a basic premise of RCM. In an effort to cut costs, for example, companies may abandon supporting system maintenance. Sometimes consequences from earlier decisions become apparent only years later. Cause-effect relationships aren't always known or established immediately. For example, switchgear-supporting heating, ventilation, and air conditioning (HVAC) systems reduce aging stress by providing cool, clean conditioned air to high voltage equipment. Many age-related switchgear failures would develop into dominant failures if supporting HVAC systems weren't maintained to remove dust and cool equipment.

Operators appreciate sophisticated water chemistry programs and their control side effects, but it has taken years to drill these points home. Companies today understand that chemistry-limit adherence supports long-term system life improvement opportunities. Painful chemistry specification lessons with high alloy steels in high-temperature water-steam solution applications have been gleaned by experience. Companies operating high-temperature steam turbines are keenly aware of water chemistry's ability to render material unsuitable for use by compromising structural integrity functions sooner. Material problems incur high costs causing immediate unavailability and capital replacement cost. Industry awareness helps companies be proactive. This costs less than learning chemistry hard-knocks alone for developing chemistry programs. Material degradation in advanced plant design high-strength alloys, and the consequent loss of functions introduced by material problems like stress corrosion cracking, premature fatigue cracking, excessive material embrittlement, and expansion problems have been the bane of many advanced plant designs introduced during the past 50 years.

The following two case studies make this point.

Case 1. In 1992, a plant elected not to shutdown after a sudden condenser tube leak. Twelve hours later,

- The condenser was scaled
- The boiler was scaled
- A boiler tube leak had occurred (from locally acidic boiling conditions)
- The unit could not be kept online due to inadequate makeup

Long-term consequences (over the next five years) included

- condenser tube leaks from complex scale-related localized pitting corrosion
- replacement of boiler water wall panel sections
- interim boiler tube leaks apparently related to boiler scale; boiler cleaning to remove scale
- boiler and condenser cleaning
- condenser re-tubing to resolve random failures from local, under-scale pitting corrosion leaks

This 350-MWe unit ran at 70% availability with a 53% capacity factor. This single event didn't cause all losses, but its secondary tube failures contributed mightily.

Case 2. Maintenance was abandoned on a difficult-to-maintain boiler ventilation system. Pneumatic louvers opened and closed louver panels summer to winter. Louvers operated off instrument air and were extremely hard to maintain due to their location. Louvers were inadequately designed and easily failed from overloading due to high winds, icing, and thermal binding. No one understood the system or its role in maintaining the boiler *as built* ambient temperatures and prescribed airflow. Over a 10-year period, 60% of the designed airflow capacity was lost, and airflow short-circuited intended summer and winter pathways. About half of the building louvers became stuck in open or closed positions randomly. Normal winter airflow—drawing warm air up through the boiler pipe chase—could not be maintained. Summer airflow was inadequate to cool the boiler burners or blowers on the upper levels, which randomly failed at higher rates from overheating.

Freezing of instrument air for igniters and startup feedwater control in the winter complemented overheating failures of flame scanners and other instrumentation in the summer. Unpredictable airflow aggravated coal dusting problems. Equipment for air that would have otherwise filtered through floor-level particle filters now entered laden with dust. Northwesterly winds in sub-zero winter conditions caused instrumentation and habitability problems. Technicians refused to work in the overheated boiler house gallery in the summer to correct randomly failed instrumentation cards. Startups and shutdowns required electricians to *jumper* failed flame scanners. Access doors at all levels were left open to improve ventilation and reduce building temperatures, thereby introducing more unfiltered air. Open doors aggravated coal dusting, resulting in an even higher rate of boiler flame scanner and instrumentation failures.

This case illustrates the law of unintended consequences. The relationship between random instrumentation failures (due to environmental stresses of dust and heat), and the cause (poor ventilation) was difficult to connect. Attempts to improve cooling by opening doors and louvers actually made overall cooling airflow worse! Resistance on the part of technicians and electricians to perform corrective maintenance was clear in hindsight, yet the connection between building environmental conditions and worker effectiveness was not. Worker cultural partiality avoiding low status, simple louver and HVAC maintenance resulted in much higher stress, high-risk maintenance accompanied by more operating risk. Lapses such as these are common and explain how some facilities fall so far behind their cost competitiveness curve that they become unprofitable to operate.

Programs not traditionally considered *maintenance* (like chemistry) have, in fact, scheduled maintenance aspects. Identifying and controlling all scheduled failure-avoidance programs with common tools, two of which are rounds and chemistry, benefit overall unit reliability regardless of the department responsible. The nuclear industry gradually learned this lesson as it replaced many aging steam generators. Pro-active care is hard to value without a high level of organizational maturity and commitment that stems from painful experience.

Component classification screens components for direct maintenance. Identification focuses maintenance efforts on component failures that matter. But non-critical equipment failures also matter. The difference is that plant managers have the luxury of dealing with non-critical failures cost-effectively as they emerge. Condition-directed maintenance based on operating rounds, predictive maintenance, and other system monitoring effectively controls these failures that don't directly *take the plant out*. To assure that this type of maintenance continues, the timely performance of on-condition maintenance is required.

The decision to defer indicated maintenance on failed equipment or to manage maintenance so loosely that failure occurs by default constitutes run-to-failure. Craft and operators express legitimate concerns about deferring indicated maintenance.

This intended meaning of the term run-to-failure contradicts the RCM-intended one, which is *no scheduled maintenance*. *Run-to-failure* does nothing—ever; *no scheduled maintenance* does something—when needed. Unfortunately in cost-cutting times, the choice may be to accept indirect non-consequential failures without correction even though their long-term effects may include:

- loss of design redundancy
- loss of alarm warning for monitored equipment developing failures
- increased probability of secondary failures
- increased dependence on operators to intuitively express otherwise alarmed, hidden failures
- decreased reliability and accuracy of critical instruments
- increased expectation of analytical insight to interpret plant operating conditions from everyone
- decreased reliability and increased cost

In short, reduced maintenance costs occur at the expense of higher operating costs to fund more operations monitoring and failure workarounds for redundant or indirectly failing equipment, increased risk of missed emerging failure identification, and lower availability. This "let it be" low-cost maintenance can only be marginally compensated with experienced staff.

Reducing short-term cost is a valid scenario when a plant is slated for near-term shutdown. For random failing items at risk in such a policy shift, the likelihood of *protected* failures occurring approaches certainty as the time interval without intervention increases. Facility unreliability grows and may eventually force a shutdown for economic or safety reasons. (Safety is almost invariably the last element to be sacrificed, for legal and cultural reasons.) Owning a plant unconsciously in this state is unacceptable. Inability to cover operating expenses, including maintenance costs, inevitably leads to poor economic operation. Case studies about industries illustrate this decline. The lax maintenance program of the railroads in the 1970s caused by low rates of return contributed to their demise and the dissolution of the U.S. Interstate Commerce Commission.

Excluding equipment that is unlikely to cause direct operating consequences and that cannot benefit from a scheduled maintenance plan is an early step in developing an RCM plan. Identifying and excluding equipment that can't cause direct operating consequences numerically reduces the master equipment list. Plant equipment coded with *tags* numerically varies from facility to facility. Equipment numbering philosophy differences make absolute value comparisons mean little; however, a large plant easily has more than 10,000 separately tagged pieces of equipment! Nuclear plants typically have more than 100,000, and one can easily argue that modern pulverized coal-fired generating stations with their environmental control systems exceed nuclear plants in complexity. Yet, critical equipment (e.g., equipment with dominant failure modes) must be maintained over the plant's operating life for economic, if not safety, reasons.

In operating plants, equipment has been identified in the CMMS/EAMS equipment tables. Having equipment tables is a godsend to an RCM development effort. Where haphazard, incomplete, or error-prone tables exist, the burden of correction commensurately slows RCM analysis progress. One less-appreciated but obvious benefit from correcting a weak equipment breakdown structure is removing errors from routine, ongoing maintenance performance. While RCM efforts incur a cost, the plant benefits.

Risk exposure development

Critical equipment classification depends upon system trains, hardware symmetry, subsystems, and skids that reflect repetitive design patterns, similarity, and symmetry. All repetition patterns simplify analysis. Symmetry and replication provide force multipliers that can reproduce and reuse developed analyses. Standardization is a principal design goal, and all RCM analysis should standardize to avoid replication, take advantage of implicit design features, and minimize analysis and database size. Equipment classification helps define the equipment partition. It precludes non-critical equipment from detailed analysis.

Identically installed equipment in symmetrical trains defines *normal models*. One function met identically by three hardware items in identical installation suggests maintenance strategy symmetry, but hardware installation symmetry alone does not guarantee complementary usage symmetry. Quite the opposite occurs regularly. Operations has long preferred equipment used for virtually identical, symmetrical installations that make equipment failure diverge from expected, identical symmetrically plans. Operations strategies can be highly effective. *Even wear* equipment rotation simply assures that otherwise redundant equipment wears out all at the same time, a situation that is not necessarily advantageous. Here it is advantageous to introduce *asymmetry* of operation to time wear-out and overhauls for different periods.

The normal model facilitates the reuse of essentially common solutions. The model captures the installation environment, service uniqueness, and equipment commonality in context. Near-identical trains, skids, or even maintenance plans for different, non-identical equipment used in essentially equivalent applications can utilize the same solution. Normal model equipment fairly represents all application cases. Normal models provide identical user reference models for reuse. Embedded models take advantage of software pointers. Using a *pointer*, software embeds an identical copy of an original, painting a screen or printing a report. Developers of a normal model can do the same thing.

In practice, CMMS/EAMS systems provide an equipment breakdown structure. Many plants use the AE-provided plant construction purchase order (PO) equipment lists for this equipment partition. Where available, they simplify and communicate. The alternative is using the available equipment list to recode the plant, which means a systematic re-numbering or otherwise identifying the plant's equipment, system by system. This should be undertaken only as a last resort. Most plant owners/operators and their engineers see no value from the exercise, and recoding equipment precludes reloading the plant's original CMMS/EAMS equipment-based WO PM plans with enhanced versions.

Equipment critical/non-critical classification

Equipment review for critical/non-critical risk exposure basis classification establishes equipment risk.

Performing equipment risk classification develops general thumb rules. Valves with remote automatic operators are normally critical, for example. Exceptions include manually activated isolation valves for nuclear post-accident monitoring or valves like heater isolation valves needed to continue operation during online maintenance. (Exceptions to these rules often identify the high-risk, non-standard equipment that benefits most from such a review.) Valves supporting online maintenance provide operational benefits. Another thumb rule is that motor/load breaker risk exposure classification is the same as that for the supplied load. The load determines switchgear importance. Using these and similar rules while knowing little or nothing about the system, one quickly reduces many systems' equipment to critical/non-critical categories. The acknowledged experts could then review the analysis and use the results to finish the risk exposure analysis and validate preliminary work in *siege session* format. One abandons the desire to deductively analyze infinitesimal detail in exchange for speed. (Does one or can one really learn the system in a two-week effort, anyhow? To what degree do system teams still fill in final results?)

In some organizations, quality assurance will object to process streamlining, questioning the validity of simplified results. Process streamlining leads to abbreviated, streamlined methods based upon empirical rules—non linear (front-to-end) engineering analysis! However, putting analytical products before reviewers faster and capturing expertise more quickly has obvious cost advantages. Achieving the ultimate objective—to help the experts become much better at understanding their own system's reliability and cost—takes investment. Without PM improvement effort, people do the same old PM in the same old way and with high costs. It's hard to argue with this streamlined reliability-centered maintenance (SRCM) case—especially if the alternative is to do no PM optimization because of its high cost.

Risk partition

Industrial facilities have hundred of thousands of components aggregated into systems to provide services that ultimately produce products. Some systems provide safety or support functions; in some plants, these are the most important capabilities. Neither the public nor workers will tolerate putting their health at risk. This intolerance is accepted modern industrial practice. Further, the public will not tolerate environmental risks according to prevailing standards. While it may look askance at an industrial accident, the public looks upon chronic safety violators or polluters as those who would compromise long-term public health.

Three risk category rankings evolved under aerospace-derived RCM: safety, production (operations), and cost. Safety—the highest rank—includes occupational, environmental, public health, and safety. Safety outranks all other risk factors. Immediate plant shutdown takes precedence over risk to worker safety. Safety outranks all other risk categories.

RCM qualifies safety risk with direct safety qualification. Direct safety consequences remove distant, multiple-chain possibilities that fault trees develop. RCM focuses upon direct safety threats to stated operating goals. Because operating goals can be stated as a desired redundancy level, this traditional limitation can be modified in practice by restating goals. Nuclear plant redundancy requirements exceed historical RCM standards for public health and safety applications. Multiple safety redundancy chains may be treated with the same risk exposure as those affected at the direct safety level.

The equipment partition breaks a facility down by system into trains, skids, subsystems, and components. Partition can use the component tag or the CMMS equipment identifier, which are usually one and the same. Most large facilities had an AE designer whose designs allow a CMMS/EAMS partition to be downloaded from the CMMS/EAMS or master equipment list (MEL) design database registry. The MEL identifies each equipment item for cost, work, accounting, and reliability management purposes.

Systems aggregately provide functions that allow facilities to produce value as output products and functions. From a system perspective, every component contributes system functionality. To evaluate how a component contributes, one must know these things:

- system functions
- equipment (components) in the system
- function contribution (from component to system)

If a boiler's functional requirement is to contain steam under all conditions of pressure—including design overpressure—and if safety valves provide overpressure protection, then safety valves support the *contain steam* function. If a generator converts rotating energy to electrical voltage and current, then excitation and windings that convert mechanical to electrical energy support generator function. While most cases are this obvious, some are not. Developing functional requirements focuses on equipment, the reason for its presence, services that are lost if the equipment is lost, and plant equipment risk.

Developing an equipment partition requires understanding the systems first. This step requires the following actions:

- Define the system.
- Define the system's functions.
- Identify its beginning and end (e.g., describing its contents and interfaces).
- Describe the way it can fail to meet operating objectives.

Most facilities' systems have been partitioned and have had functional descriptions developed. AEs develop original design documents based on functional requirements. P&ID drawings, takeoff lists, and purchase specifications for vendors who supply the equipment develop system partitions. Equipment procured becomes part of the system. Secondary drawings like plan drawings, electrical one-liners, and control drawings amplify basic design requirements summarized by the PID drawings. All these materials are developed as a part of the original design specifications and stay with a plant. They need not be reconstructed unless they are lost or hopelessly out of date.

Streamlined RCM Steps

Streamlined RCM can be summarized as follows:

- Download the master equipment list.
- Classify systems using risk, production or cost criteria RCM analysis.
- Pick a system(s) for analysis.
- Identify the system's functions.
- Identify the system boundary scope and interfaces.
- Locate and prepare reference materials: P&ID drawings (critical equipment) and design descriptions (for functions).

Why systems?

Designs start with systems—function-oriented, high-level design descriptions. Large facility design is based on systems. If one understands systems, the designer's building blocks, one understands the design. Systems partition functions so that designs may be formalized and translated into equipment construction requirements.

Why functions?

The easiest way to evaluate equipment loss or, equivalently, identify the contribution to system functionality, is to discover the equipment that can cause functional loss. With functions restated as failures, it is simple to picture the components that cause function loss and the way any component failure affects system functions. For example, if a pressure vessel function is to contain fluid contents under all conditions of pressure and overpressure, the function failure statement is *fails to contain contents under all conditions of pressure and overpressure*. This statement enables the designer to

- focus on equipment providing protection
- establish design limits for credible overpressure
- identify the catastrophic secondary failure (vessel overpressure rupture) protected
- think inductively about how higher level functions that could be affected by component functional losses

Functionality flows upward. One view is that systems are a hierarchy of supplied functions. In integrating equipment into systems, the challenge is to avoid introducing functions that are not required and that would introduce unknown or unexpected system functional failures. Doing this well is a matter of expert design, learning, and conscious effort. Since systems must be integrated from hardware, introducing undesired functions and their failures is always a design risk.

Picking systems requires consideration of risks and consequences that vary by systems. Instrument air loss poses less direct failure risk than the steam bypass system's failure, for example. Safety, production, and cost provide the natural ranking upon which to select systems by risks for analysis. Some systems are resource-intensive by design. Pulverized coal primary fuel systems and coal-handling systems fit this category. Part wear; dusty, high-temperature environments; and acidic moisture introduce rapid aging and environmental stresses that accelerate failure. Other systems require little maintenance. Generator hydrogen supply demonstrates high reliability with very little support in spite of the complexity involved. Designs that have matured—they have been around for 30 or more years—have fewer problems compared to new designs. Problems in mature designs are more manageable and pose fewer operational and safety risks. Over time mature design weaknesses have become well understood. Inherent design limitations have become apparent with age. Ultimately, superior designs outflank mature ones by the advances in technology.

Once a system is selected, system equipment can be identified from PIDs downloaded from the CMMS/EAMS, and risk exposure assessment can begin. All equipment in a system supports that system's functions. Theoretically, no equipment lacks useful function; all equipment has a meaningful purpose in design.

Practically, design is inexact. Thumb rules, culture, and even seat-of-the-pants methods fill out designs. Customizing in the North American generation market means that although standards are evident, most designs have moderate degrees of customization. Some designs evolve on the floor or during construction on the fly.

Old plants have a rich legacy of modifications, some to many of which may not be reflected in designs, based on industry. Furthermore, even design group standards and sometimes-inapplicable standards add equipment *a la mode*.

One past standard called for installing check valves for all instrument air lines. All had moisture drains that required one-way check valves for plant and instrument air. Most of these valves rusted and bound up where moisture was present. The original design intent was to provide adequately for drainage. The outcome failed in this regard. Other equipment subsystems suffer the same problems.

One Powder River Basin (PRB)-fired coal mill installation had five fire suppression systems. The most endearing of the lot was a soap surfactant injector system the operators affectionately called *Mr. Bubbles*. The four other systems were

- steam suppression
- condensate spray suppression
- firewater suppression
- chemical injection

As might be predicted, the underlying problem—primary fuel unreliability, including coal mill fires—persisted in spite of all these fire suppression systems. The reliability of the automatic suppression systems was low, and operators used water suppression to extinguish blazes, quenching hot metal parts and diminishing their life. Reactive maintenance is better focused tackling root causes. In the intervening years, since observing this multiple coal mill PRB fire-suppression installation, many companies have mastered the PRB fuel combustibility control.

Plant support engineering may seek design solutions to operational problems that could have simple maintenance or operational solution methods. The temptation with any problem is seeing only one way to solve it—not the simplest way. (If all you have is a hammer, everything is a nail!)

Why identify risk exposure?

An effective method for identifying system equipment risk exposure while partitioning is to use a set of color system PIDs with highlighters to represent categories visually:

- Red—safety
- Yellow—operations (production)
- Green—cost

PIDs are systematically reviewed, marking each equipment item with its appropriate color based upon assigned category. Knowledgeable plant engineers can rank system equipment in two (or fewer) days and with 80–90% accuracy. To get to the 95%+ range, operators, mechanics, and technicians must work as a team to review and to further refine details. With its single failure assumption, RCM plans effectively deal with single failures. This restriction makes analytical results look limited—particularly at plants with repetitive multiple failure events or application environments where users can develop fault trees. Although these more complex analyses' resulting plans have little practical utility, they illustrate one notable point: No amount of RCM overcomes the fundamental absence of a maintenance program! Multiple failures will eventually cause system function losses; complex multiple failure chains are primarily responsible for the system level failures that occur in everyday plant operations practice. RCM provides simple answers to preclude complex events—when a maintenance program is in place.

An efficient maintenance plan does not eliminate maintenance altogether. Conversely, an efficient plan makes maintenance timing more sensitive than ever! RCM precedes failure with maintenance performance. This push for maintenance is intrinsically incompatible with reactive maintenance processes. These environments often lack the sophistication to work condition-directed maintenance.

Equipment identification systems support critical equipment risk exposure lists. Users are forewarned: critical risk exposure assignment at the hardware level implies equipment supports one functional category, prioritized by the equipment under consideration. In fact, most equipment supports more than one function. For example, a turbine steam extraction line typically improves heat rate. Extraction lines contain check valves. In a perfect world, extraction lines bleed steam off operating units' turbines. Practically, during transients (like turbine trips) heaters flash and backflow condensed steam into the unloaded turbine potentially causing acceleration to destruction—without check valves.

As steam extraction was historically introduced, design installation of extraction bleeder/check valves followed based upon this expensive lesson learned. Today, insurance companies require check valves, largely reflecting the lessons of this hitherto unrecognized equipment risk—turbine overspeed runaway under no-load conditions—introduced with extraction steam designs. Extraction steam lines must (1) provide extraction steam under normal operating conditions for heat rate improvement, while (2) isolating potential reverse flows under transient conditions with low or no turbine loads. Simplified component representation as one risk category merges real equipment multiple-dimensional functions, failures, and risk into one classification. Simplified analysis combines functions and functional failures into composites. Though superficially viewing equipment simply—critical or not—composite equipment risk representation obscures real risks derived from different equipment failure modes. These real risk drivers are what RCM seeks to reveal and avoid.

So for simplicity, equipment worst-cause dominant failure mode criticality assigns a corresponding conservative component criticality designation. This provides a simple presentation format. This approach is simple and elegant—if one can remember that the component's criticality is driven to any level of SOC by its worst case dominant failure modes. These can bias the failure partition analysis if analysts forget that component failures—not intrinsic criticality assignment—pose actual risk.

This solution, component criticality assignment, implies all component functions/failure modes have the same criticality, and they generally do not unless these are specifically displayed with a failure-level criticality. In fact, rarely will even two different functions support the same criticality role, or indeed necessarily both even be critical.

Criticality can be ranked into three primary levels with traditional RCM analysis. Since SOC depends on the failure mode—the most restrictive mode—the worst dominant failure mode must be used. In summary, when identifying dominant failure risk, consider

- criticality
- risk exposure subclassification SOC
- component assignment to highest partition risk
- dominant failure mode(s)

Failure Modes and Effects Analysis

Failure mode describes how something fails. Failure mode effects describe failure local effects that provide evidence of failure at the component level (*see* Fig. 2–15).

With the black box equipment model, given that all inputs are present, failure to provide output—the desired function—must reflect an internal fault (failure inside the black box). Failure must occur *inside* the hot model, or one or more inputs, using a functional block diagram construction, must be missing. Reliability engineering models failure. Failure must be adequately defined to address with scheduled maintenance.

Failure characterization and development is part engineering, part art. Failure modes describe how components fail. Components are made of parts. Part failures cause component failure. Part failure is conventionally described as an engineering failure mechanism. Parts are the lowest discernible element in the equipment hardware hierarchy chains. The part is the atom of the equipment—the lowest detail level. Failed parts are the elements maintenance activities must identify and correct. *Rework or replace* addresses failed component parts. Replacing or reworking parts corrects failures.

This hardware hierarchy (*see* Fig. 2–16) and failure description can be summarized this way. It

- differentiates components into parts
- details components failure effects as failure modes

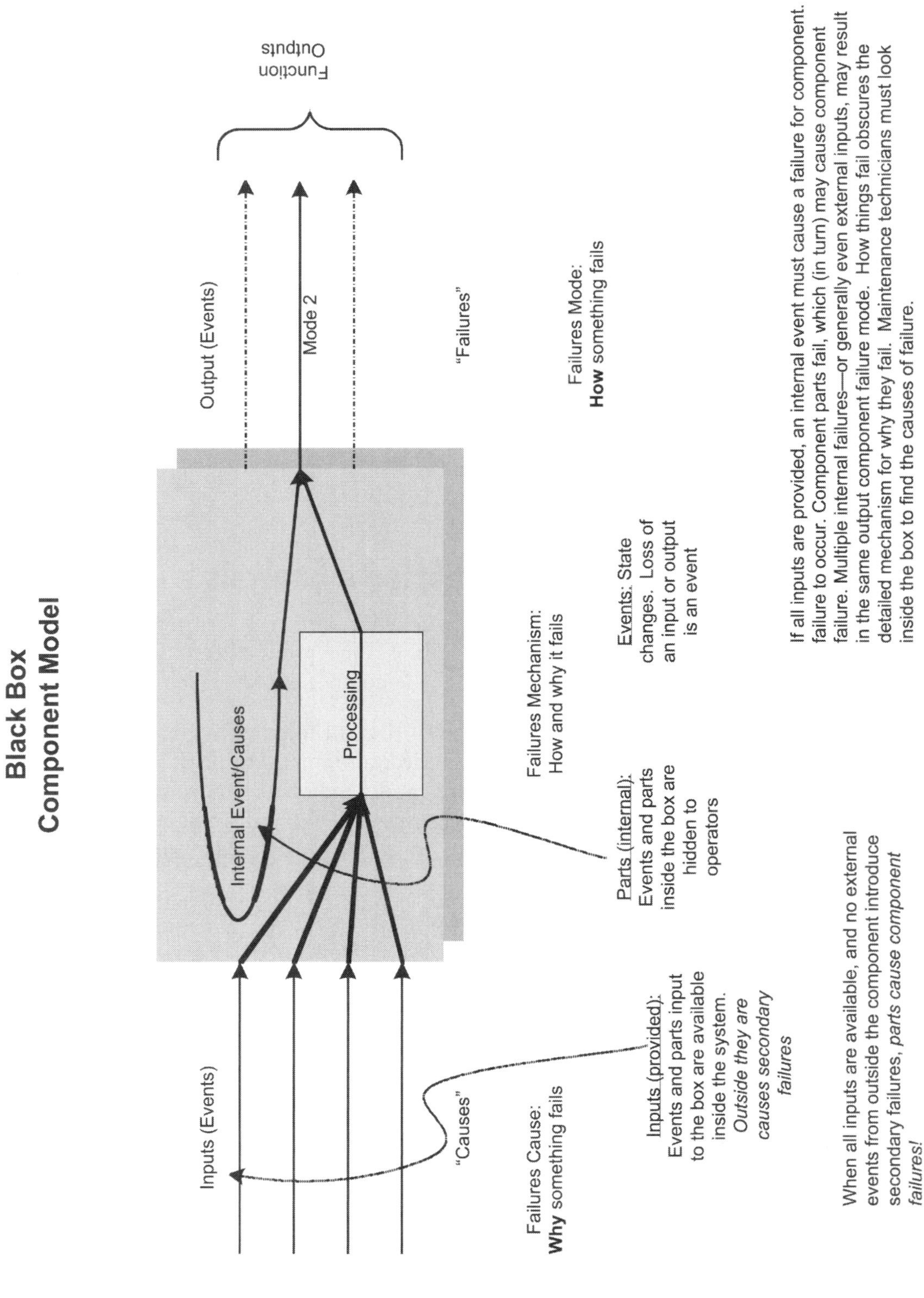

Fig. 2–15 Black Box Component Model

- describes part failure for analysis
- explains part failure completely
- provides
  - failure (mode)
  - cause
  - symptom
  - local effects
  - hidden failure (or evident)
  - risk exposure

Failures may or may not be evident; often they are hidden. Symptoms relate how a failure presents itself—and it eventually must become evident—and suggest how to find failure at its onset. Local effects explain how the failure propagates, and risk ranks the failure consequences in standard SOC criteria.

For example, contact points in a medium (4-13kV) voltage breaker bus include stabs, primary, and secondary arcing contacts. Consider secondary contacts. These can erode and burn from normal use. Consumed on use, they are replaced, preserving the

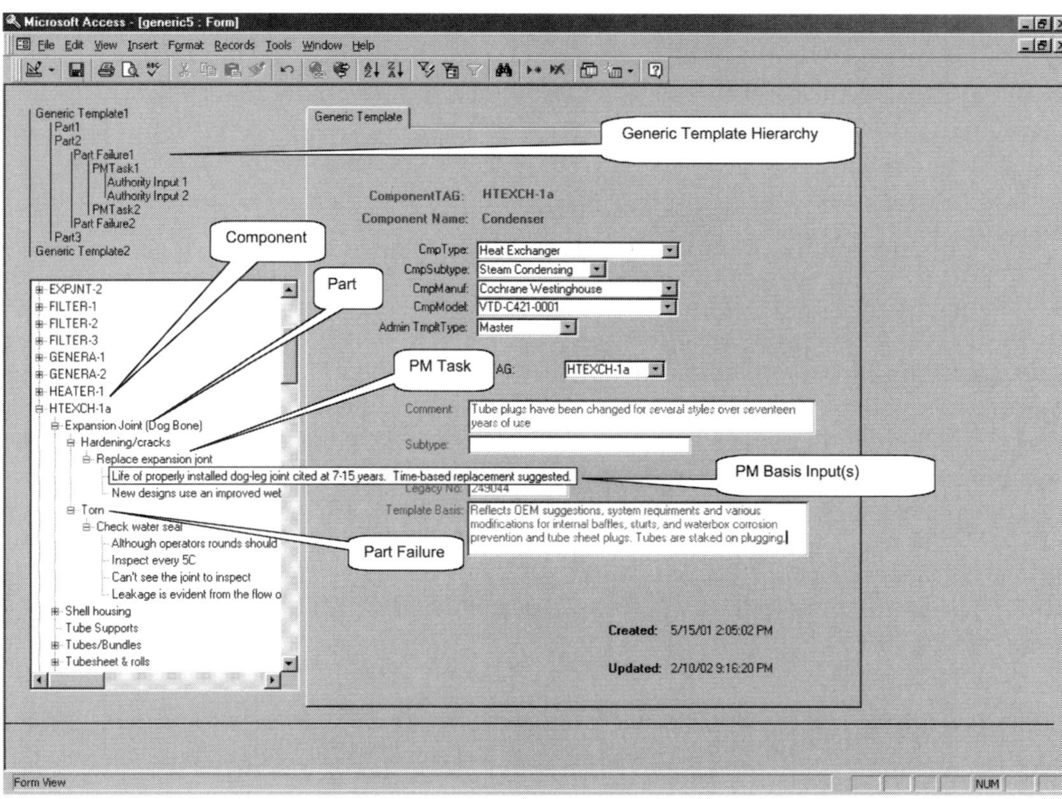

Fig. 2–16 Hardware Hierarchy

main contacts. Symptoms of material erosion aging include erosion loss or excessive pitting on the main contacts, which are hidden. (An operator can't see this.) The normal risk of eroded arcing contacts is more arcing on the main contacts, thus increasing their pitting erosion. All contacts require a certain amount of contact pressure to conduct without heating. Too much pitting erosion and heating occurs. At very extreme limits, overheating could cause tripping from auxiliary contact relay protection resulting in a breaker failure.

In component-part hierarchy failure terms, the main content ranking is as follows:

- [component] – breaker
- [component failure mode] – fails to carry load
- [part] – main contact
- [part failure] – pitting erosion resistance
- [symptom] – high temperature (hidden)
- [local effect] – higher temperatures; increased pitting metal loss; eventually won't carry load
- [risk] – operational

Note that the operational assessment hinges on other operational components or parts performing their job well. In contrast, the arcing contacts component failure hierarchy has this ranking:

- [component] – breaker
- [component failure mode] – fails to preserve main controls
- [part] – arcing contact
- [part failure] – erosion
- [symptoms] – high temperature (hidden)
  – pitted main contact (hidden)
  – longer operational travel (hidden)
- [local effect] – main contact pitting
- [risk] – cost

Risk for the arcing contacts is primarily *cost*. As the arcing contacts fail, the main contacts arc more, age more, and wear out faster. Also note that tests can reveal deterioration, such as longer contact operating times and indistinct voltage drop or rise across contacts during operation.

Partition detail level

CMMS/EAMS equipment lists suffer from two general problems—too much and too little detail. Too much detail stems from excessive equipment coding. Technicians coded plants, armed with regulatory guidelines. Some did little coding, others a lot. Some numbered virtually all conceivably identifiable plant hardware items. As assembled equipment was examined, numbers and details expanded. A single nuclear unit often has more than 100,000 coded items.

Two strategies can address too much detail: *grouping* and *making non-critical.* Grouping inverts coding. Grouping components subverts detail into functional subsystems, skids, and loops (control loops) and aggregates components that should not receive individual scheduled maintenance attention. Associated components (electronic component loops, for example) don't receive individual attention until failure occurs. A single equipment tag provides a nominal PM task and target, which is probably a calibration and channel check. Grouping places associated equipment under its *primary tag* (*see* Fig. 2–17). Grouping (or making a component non-critical) removes associated equipment from active consideration. Assessment separates the trivial many from the important few.

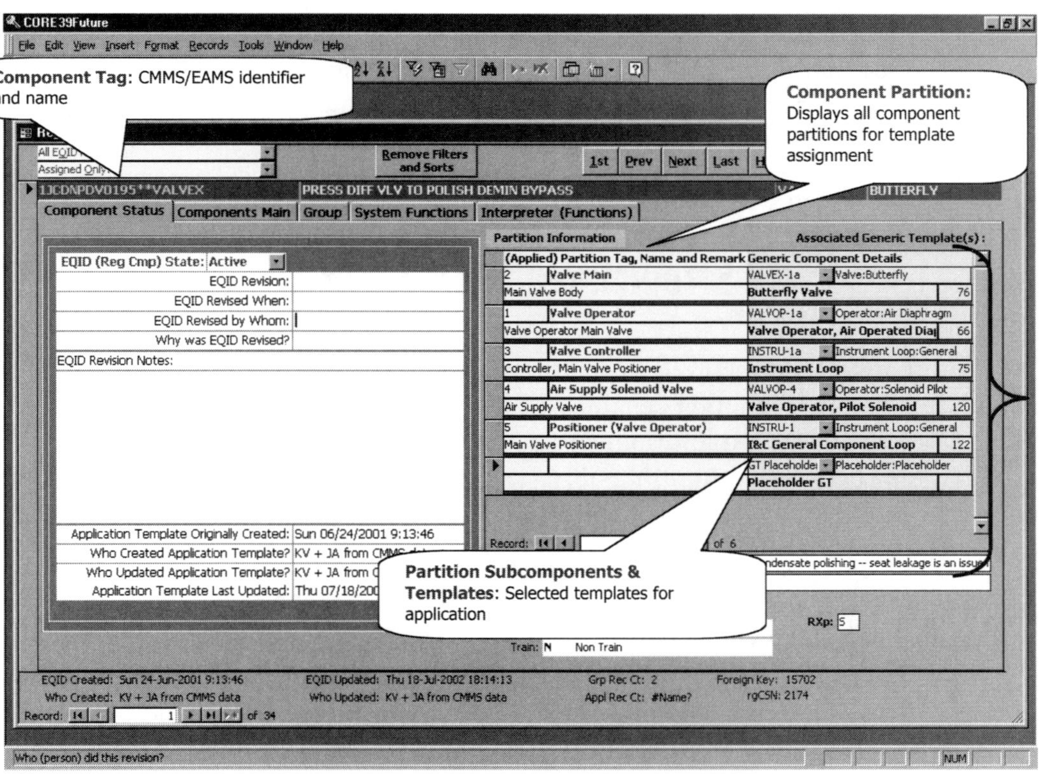

Fig. 2–17 Component Partitioning for Multi-Template Assignment

Providing the basis for equipment non-critical status helps with audits and maintains results. RCM thumb rules consider design, but operating staffs rarely do so consciously. Consider non-critical manual valves. Functionally, manual isolation valves usually facilitate maintenance. Performing maintenance in the absence of safety or operating consequences is based upon cost. Two valve failure modes dominate—seat and stem packing leaks. On-condition maintenance can effectively deal with either. The valves assure on-line isolation to perform on-line maintenance or deal with operating event isolation. *Valve stroking* wipes crud off valve seating surfaces, relaxes stem packing, and restores packing lubrication and valve-to-stem contact. These actions extend valve life. Operating organizations are vaguely aware that manual valves need little maintenance. They're unaware of the value of valve stroking, and most have never thought about manual valve maintenance strategies.

Fossil plants may lack equipment list detail. Too little coded equipment can be addressed by adding equipment item-by-item to CMMS/EAMS tables. New equipment lists can also be prepared and imported into the CMMS/EAMS. Importing is easier when many tags are needed. Other alternatives are to add new CMMS/EAMS equipment tags with RCM software, building an equipment subcomponents list within the software. In this latter case, the supported CMMS/EAMS system import capabilities should be checked first to avoid the unpleasant surprise of not being able to use developed work.

Primary/secondary associations employ logical pointers to use previously developed analysis. These embedded links replicate familiar Windows link editing. They convey data directly, replacing links with actual values. Reduced database size and quick update capability results.

Streamlined RCM Justified

Traditional RCM evolved from the airline industry in the age of mainframe computer systems. Computers enabled United Airlines to study early part failures and usage, leading to age exploration. From these and other studies, Nowlan and Heap disputed the cherished notion that all parts age. Nowlan and Heap worked before the advent of PCs and on large fleets of nearly identical airplanes with outstanding parts and work order records. Few companies maintain such detailed records outside the aerospace industry. Yet today,

- PCs are prevalent.
- Analysis extends to other engineering and support fields.
- Operating costs and public risks worries predominate analysis.
- More groups and people do analysis.
- More experience is available.

These factors ease RCM application in large installations today.

RCM standards and variations abound, while traditionalists advocate classic *seven-step* RCM. Practically, the best *shelf* RCM analysis pales beside imperfectly implemented analysis. Implementation is the practical goal of all RCM and the primary industrial operating companies' barrier in process, power, and electrical transmission and distribution industries. Unless implementation can be made easier, RCM costs will remain high. Theoretical correctness, tempered for practical acceptance, helps realize implementation. Analyses need not be perfect; they follow an 80/20 rule of *effectively correct, readily implemented*, in a flexible program that easily incorporates learning. Potential for error predominantly involves cost and operational impacts. Partitioning to identify safety and operational risks assures unacceptable consequences are never missed. *Living maintenance* improves information flow and discovers new opportunities. These extend the paradigm so that plant maintenance engineers, managers and workers can achieve practical maintenance response that is timely, accepted, and easily tuned with feedback for little additional effort. The need for perfect analysis is tempered with the ability of the program to adjust and correct to follow equipment reliability requirements.

The best programs grow by drawing more participants into the fold. More users improve insights, which changes analysis. But this growth curve contains its own Catch–22: Its success forces rework as more users critically review previous analytical results with their new perspectives and experience.

Implementation has practical barriers, of which one is the need to re-enter CMMS/EAMS data. This discourages the users of older CMMS/EAMS that lacks interfaces and word processors that speed editing efforts. Some administrative systems require that all PM changes be justified; this adds burden. Burdens placed on data entry people practically discourage work performance. Traditional RCM and its documentation burdens discourage analysis performance, discrediting the RCM process. Successful RCM has to achieve field application; an implemented process realizes practical success. Practical implementation has been achieved by removing the grunt-work burden from the end user. No one today performs mathematical integration because the process is embedded in every modern digital controller. Streamlining RCM offers the same benefits—practical success though practical field implementation.

SAE JA-1011, RCM Processes, provides a standard by which to judge whether processes are *RCM processes*. While this effort is laudable, some JA-1011 passages cause concern among other professional reliability working groups. One clause implies that RCM processes must be linear. Plant engineers with many years of experience in complex facilities find that in living programs, PM enhancements need not. SAE JA-1011 represents an idealized engineering view, which provides a useful process benchmark. The best processes are sustainable on the shop floor. JA-1011's idealized flows do not (in the author's opinion) support practical RCM floor implementation in the ordinary ebb and flow of maintenance work amidst operations, outages, and the emerging cacophony of failures that demand a plant engineer's attention. It does provide a process suitable for project work with repetitive analysis with a core group. Real-time RCM is driven upward, and its processes can be tuned (with learning) to provide outstanding results once users learn the process.

Streamlining RCM techniques: templates

Streamlining techniques accelerate and standardize RCM analysis. PM programs provide *applicable and effective* scheduled maintenance tasks. Applying RCM criteria eliminates non-justifiable tasks quickly. These include tasks with no exhibited failure or those that lack craft validation. An existing PM program's plan provides a useful step-off point. The primary objection to their use is that it biases results. It would not follow traditional RCM processes. When analysts know existing scheduled maintenance plans, they become biased, compromising analysis. (Using an existing PM program also smacks of cribbing, and who wants to cheat?) On the other hand, starting analysis from basics over and over for each new analysis slows RCM and increases cost. Speed and job completion interests suggest compromise. Documenting all tasks' failure prevention applicable/effectiveness basis, using rigorous criteria, assures valid analysis.

Templates streamline technique. Vendor PMs address common dominant failure modes. These are a form of template. Using every failure mode ever expressed for every PM work plan on equipment leads to complex programs that enumerate hypothetical failures that are out of touch with plant-dominant failures and operating reality. Template task enumeration carried to the extreme becomes PM optimization (PMO). PMO applies tasks without considering risk, which must be avoided to preserve RCM. PMO applies the same generic tasks to all components of a given component type, ignoring criticality risk, service use, and environment. On the other hand, extending generic template-based PMO to capture risk context achieves RCM. Generic templates can provide a rigorous dominant failure mode basis, providing tasks can be applied selectively. The engineer's temptation is to indiscriminately select failures and their prevention tasks, exceeding those expressed in a specific case, for *completeness*. Theoretical, one-time failure modes from reference sources, applied unchecked, bog down a site program with *analysis paralysis*, obscuring real site failure experience. Rare, improbable event analysis causes ineffective scheduled maintenance plans and resource use. The trivial many dilute the critical few. Use of rigorous failure data ensures that meaningful tasks are selected avoiding this trap.

Exposure risk classification allows risk-based template application. Applying templates to installed components reduces users to deciding which failure modes apply, then applying them while addressing usage environment context to adjust time interval. Strategies must accommodate variations in service, environment, and other usage factors that affect aging, predominately expressing specific failure modes.

As RCM teams learn, they rework earlier work. Managing rework requires measuring rework statistically. Rework that reflects learning is not a liability; rework to reformat or correct oversights, omissions, or other avoidable errors is undesirable. Formatting, locking, unlocking, blocking, and re-blocking tasks are time-intensive administrative chores. These should decline as any project's processes mature. Finding and understanding rework develops a consistent process that is under control. Rework can become an end in itself and, therefore, needs careful management.

Streamlining the RCM process

Learning RCM requires a paradigm shift, and like any process, change requires formidable effort. Once operating under the RCM process, emphasis should shift to performance efficiency and speed. Working faster requires conscious effort. Calculated, *fast-track*, streamlined RCM analysis methods can be attained. The Electric Power Research Institute (EPRI), vendors, and various RCM process experts have developed streamlined methods. Conversely, at least one equally qualified, respectable authority has argued, with justification, that streamlining RCM is penny-wise and pound-foolish. Radical streamlining must be undertaken seriously. Methods should be carefully understood and benchmarked against best practices. Once selected and developed, challenging organizational processes is tough. Legacy processes always pose a barrier to RCM or any other new technology.

Thoroughness, quality, and cost paradigms run deep and express a profound engineering culture. Entrenched process changes in regulated environments are viewed skeptically. Analysis details affect quality, cost, validity of results, and other tangible RCM dimensions. Streamlined RCM critics express legitimate concerns. In the nuclear environment, RCM documentation and process requirements have made RCM uneconomical. Here, PMO is pervasive because it is closer to ad hoc shop floor practices than engineering-oriented RCM and can better distance as a less rigorous PM development process. Unlike RCM, PMO doesn't require critical or dominant failure connection, so non-experts can perform analysis. Others can't question the quality of results. Understanding cultural philosophies helps select the best RCM methods to apply. Philosophy should be considered before embarking on RCM-based PM development.

Regulators invariably found some approaches unsatisfactory in the past. Whether voiced or not, their opinions weighed heavily on RCM acceptance. RCM, as an exact, risk-based process currently aligns well with risk-based nuclear regulatory practices.

Successful RCM offers a living maintenance program. Rationalized, streamlined RCM best manages overall cost and reliability interests in many organizations. An RCM-based program without complementary maintenance ownership makes RCM success less likely. The potential benefits in most industries are so strong they are compelling.

Systems and Functions

Systems understanding

Learning complex technical systems from the ground-up takes time. Maintenance and plant systems engineers must learn fuel, turbines, control rod drives, refueling equipment, equipment cooling, and many other systems plus their many nuances.

Even after years, many feel they still have much to learn! One person cannot learn a complex system in two weeks, yet RCM analysts must do exactly this! Time is better spent learning a system only well enough to discover its reliability issues for presentation to responsible owners (usually the systems and component engineers and their systems support teams). Presenting system equipment owners the weaknesses of their equipment strategies for expert review stimulates their thoughts. Their knowledge and experience provide the major RCM process benefit concurrent with the maintenance plan changes. Documented outcomes result. Developing system questions and issues captures system requirements supporting new or modified PM tasks (*see* Fig. 2–18).

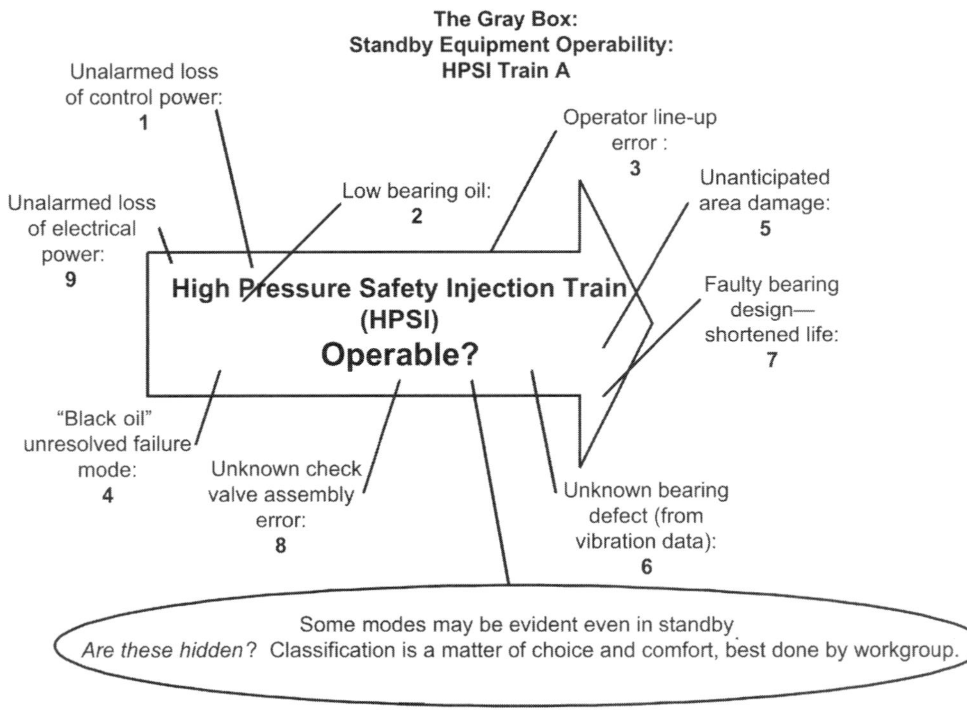

Fig. 2–18 Failure: The Gray Box

System partitioning

Breaking down facility constituent systems retraces designer steps. Systems provide design functionality to meet production, safety, and cost requirements. Differentiation into systems, sub-systems, and equipment is a first analysis step. AE system descriptions from startup provide source material. For older facilities, design may have evolved where system design descriptions may not be directly available. Old designs may also require update. An RCM effort may have to reconstruct the formal, expressed intent of a system's design.

Systems implicitly develop facility design requirements. (*see* Fig. 2–19) They include vendor requirements that may not have been provided explicitly with the plant. Turbines, for example, must meet ASME and insurance requirements. Logic schemes must pass IEEE protection logic standards. These requirements support process flow diagrams for turbine-supporting equipment such as controls and protective devices. The design documents may not be available—perhaps they were never purchased or were lost or destroyed after startup. (One plant lost virtually every design document in a 500-year flood!) For those old enough to remember, plant startup testing formerly consisted of functionality tests that ensured the design delivered the owner's contracted plant needs.

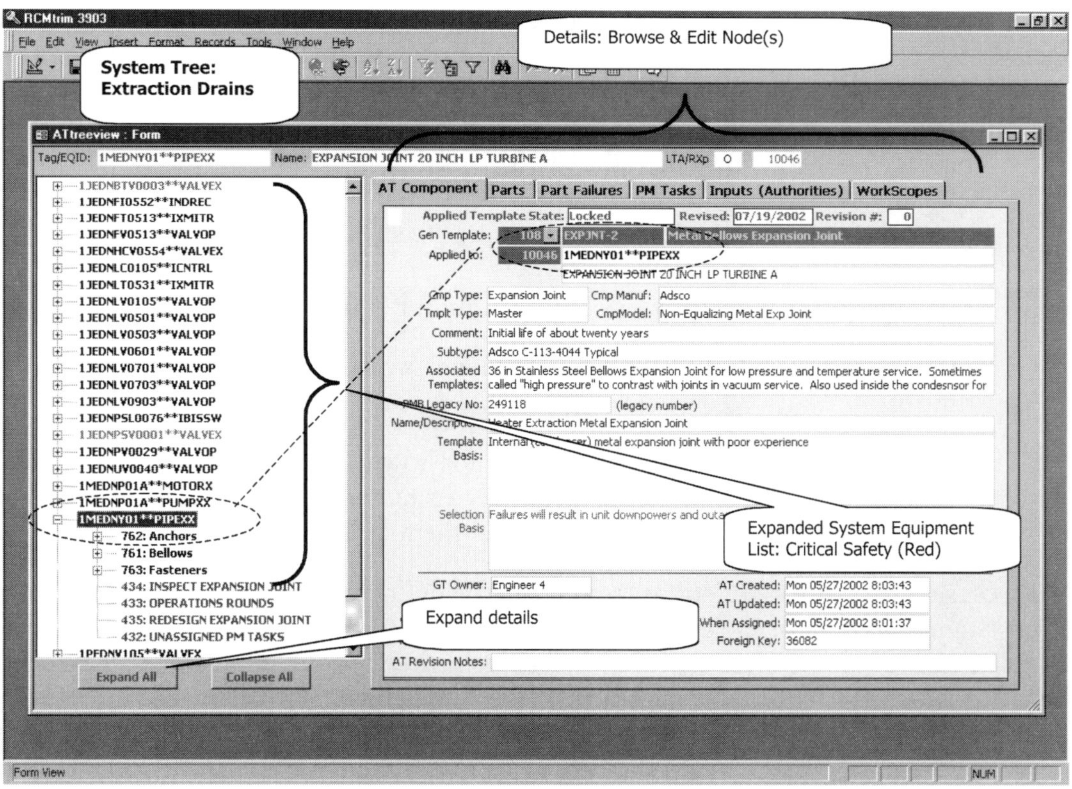

Fig. 2–19 System Tree for Expanded System Equipment List (Critical Safety)

System equipment partition differentiates trains, skids, and components. The partition identifies system functionality, from systems descriptions and functional block diagrams, redundant features, major service, and interfaces. Interfaces can be incidental boundaries with other systems or the exterior environment. Significant interfaces identify where other systems' inputs are provided, as well as the systems external inputs. Interface failures compromise system outputs. Providing for services could create a minor service interface or a key boundary. For the extraction drain (ED) system, steam and feedwater heater tube wall boundaries define a main pressure boundary. Heater tube leaks co-opt separation fundamental to either system.

Functions in documentation

Functional statements express system functions. Explicit statements are more useful than inexact ones. Exact statements capture specifications that may require additional information. Explicit functional statements define functional failures. This captures system design. System descriptions describe designer-intended functionality at plant start-up. System functional statements specify plant needs. Consequently, system functional statements specify requirements, not how they are provided. They don't include equipment. (Equipment provides the requirements that meet the specifications.) System specifications could be the following:

- provide overpressure relief above 2000 psig
- provide 2.3 million gallons per hour total feedwater suction flow
- regulate output power to less than 2%/minute rate of change
- provide full flow steam bypass at power levels below 50% thermal rated output

These specifications describe requirements, not how the requirements are provided. Based on experience, engineers think intuitively in terms of hardware such as safety valves. Engineers interpret specifications and fill them.

Occasionally, when functions aren't clearly defined, designers make inferences from similar designs or experience, without reducing those assumptions to explicit criteria. As little discoveries captured by analysis, they become nuggets of RCM. It is hard to identify requirements never clearly specified. In older facilities with evolutionary designs, one must reconstitute designs. Designs and design intention can be lost over time through no specific cause other than their very success. Good designs reflect quality construction and require only modest, familiar effort to maintain. These are taken for granted.

As plants age, hitherto unexpressed failure modes emerge—particularly as secondary or composite failures. At this time, reconstructing design functional intent becomes challenging. Validity of the failure mode is clear only in hindsight. Although the plant has never seen a failure, arguably failure could be built into the plant design. Some failures must never occur, and so new failure occurrence should never introduce failures that could have been foreseen by benchmark or any other means.

Function restatement

A functional failure statement restates a design-required function as a failure. Restatement prepares for the next step—identifying the equipment that can conceivably cause function loss (Fig. 2–20). From the functional requirement examples, one can derive the following failure statements:

- fails to provide overpressure relief above 2000 psig
- fails to provide 2.3 million gallons per hour total feedwater suction flow
- fails to output power to less than 2%/minute rate of change
- fails to provide full flow steam bypass at power levels below 50% thermal rated output

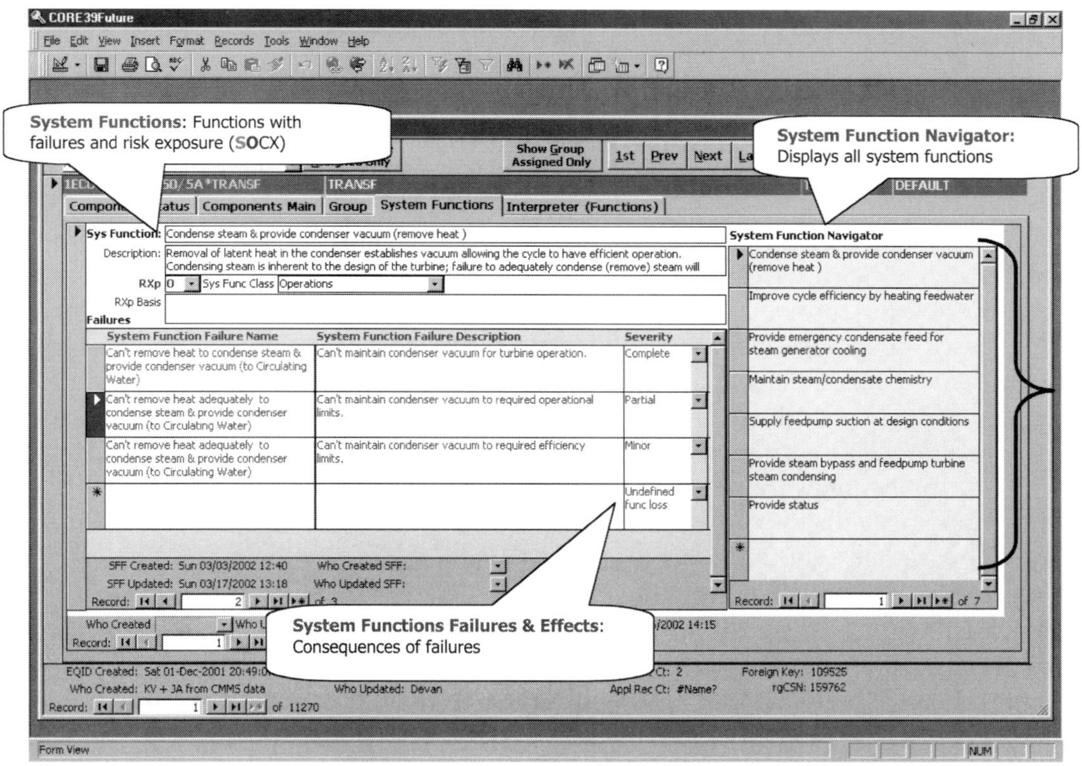

Fig. 2–20 System Functions and Function Failures: Condensate

Functional requirements

Restating functional requirements is a simple grammar exercise. Where design documents are out of date, functional requirement specifications may need to be redeveloped, or *reconstituted*. Functional requirements can be found from critical equipment WOs. To document safety relief valves limits under test, engineers should look at the responsible safety valve(s) that provide the relief function, their setpoints, design flow rates, and proceed from there. Where equipment performance specifications are not available, engineers may want to find the associated WOs. Reanalysis development is a last resort.

Checking functional requirements that have been long lost introduces new work into an otherwise stable maintenance environment. Plant maintenance staffers may react negatively. The motivation to do this work is lacking in the first place. Motivation comes from understanding the avoided consequence costs of the failure that could be present. Several years back it was discovered that several feedwater heater safety valves had never been lift-off tested in several 1959 vintage, 100-MWe units. Upon removal and preparation to test, it was further discovered the valves had rust-plugged from years of unlifted service. Clearly, this was a safety concern as well as a program oversight.

Components

Components provide functionality. In perfect designs, components provide functionality efficiently. Conversely, in new, one-of-a-kind pilot designs, equipment may never cleanly provide the functionality sought. Equipment may be abandoned in place, based upon cost consequence or design inadequacy. Often, no one knows why the equipment functions were needed in the first place! Some of the most interesting and rewarding project discoveries involve abandoned equipment that played important cost, operational, or even safety roles that evidently were lost over time. Plant operators were sometimes unaware of these roles. Some discoveries involved monitoring and alarm equipment that was difficult to maintain, so it fell out of service. This may be caused by simple ignorance, intense demands of high-maintenance equipment, or installation issues. Installations that impede maintenance contribute. Plant environments that are hard on hardware are inevitably harder on people. People are habitability barometers; they avoid working in inhospitable places, whether scheduled or correction maintenance. Maintaining environmental HVAC, lighting, access (elevators and stairs), habitability and other services or equipment often enables work. Less than ideal plant areas still have maintenance needs.

Sometimes, the maintenance shop makes a unilateral design decision to abandon service. Occasionally decisions fail to receive proper review. In some cases higher-level policy would have changed that decision; in others higher level policy would have become aligned with these working level choices. These include decisions to perform or not perform critical maintenance. Because maintenance is out of sight and mind in large companies, anomalies can be lost. RCM projects identify and incorporate missing links in a maintenance program. As RCM project "finds," these oversights of commission and omission have been responsible for substantial RCM system improvements. The *un*event is intrinsically hard to find, harder still to value and retain in the collective consciousness—thank heaven! Precursor event-avoided failures quickly fade into obscurity. Program successes may be taken for granted unless the basis for work selection is maintained.

Design functionality sources

Designers put equipment in systems to provide functionality. While some systems' primary requirements stem directly from the facility output or product, other systems provide services. Environmental systems provide cleanup and monitoring. Fuel systems provide combustion, reaction, or supporting fuel requirements. Cooling systems remove waste heat. Fire systems detect or suppress fire. Heat and ventilation systems provide habitable equipment environments. The more efficiently any functionality is provided, the less capital expense is required. Up-front cost for redundancy provisions, on-line maintenance isolation, and other design features that allow and mitigate equipment operational losses influence long-term facility operating costs. Consequently, all equipment in a system originally was included because designers thought it cost-effectively supported operating functional requirements.

Systems that have been in commercial operation for five or more years pose low risk. With new plant designs and technology, however, risk increases. Looking back 100 years at power generation design advances, some quantum leaps include:

- condensing turbines
- feedwater heating
- extraction steam
- reheat cycles
- supercritical boilers
- pulverizing attrition coal mills
- nuclear plant designs
- combined-cycle plants and combustion turbines (CT)

New advances often met profound surprises. (What technology on this list isn't commonly accepted today?) Most power generation people remember supercritical boiler metal problems in the early 1960s. Design temperatures and pressures plateaued as a result. In the 1930s, new steam extraction turbines failed from overspeed after plant trips. Steam backflow through unloaded turbines and their accompanying destructive centrifugal forces had not been anticipated. Dramatic plant size and capacity increases resulted from pulverized coal mill primary fuel systems, to only meet dismal financial losses in the 1970s from inflation and delayed production schedules. Along the way, many supporting designs were drafted, tried, accepted, or in some cases relegated to the ashbin.

Over 40 years, combustion flue gas cleanup technology has evolved from wet scrubbers to precipitator, fabric filter flue gas dust removal, and then back to dry scrubbers. Nuclear power superseded coal as the most advanced design, receded into the background, and now competes again. CT combine- cycle gas plants have achieved commercial success only to precipitously decline over the past two years. Environmental re-regulation, Kyoto accords, and Middle Eastern conflicts may yet place nuclear back on top through an equally unpredictable new set of circumstances. Technology never pauses.

Supporting equipment has likewise evolved. Reverse osmosis superseded demineralization makeup water treatment. On-line water treatment includes full-flow demineralization. Wastewater treatment includes wastewater concentrators to extend water supplies and supplement evaporation ponds. Zero-discharge waste water permitting has been common in the West for decades. Economic fuel features such as PRB coal blending were introduced then eliminated. (Mining companies do this today at much lower cost, and expensive coal blending facilities are idled or simply provide storage.)

Some new technologies fell by the wayside; others advanced. Early digital valves posed problems. Powder River Basin coal made dust suppression functional demands skyrocket. (Most early dust suppression systems were grossly undersized for PRB coal.) Power savings performance improvement made variable-speed induction and forced-draft fan drives economically desirable.

Nuclear generation experienced similar trends. Marginally economic sites were shutdown. Early oversized radioactive waste or *radwaste* facilities were scaled back (a rare occurrence!). Emergency systems were augmented, and, after the Three Mile Island (TMI) core damage and radioactive gas release event in 1979, design reassessments were pervasive for a decade. Nuclear plants saw intense regulatory and quasi-regulatory body interest in maintenance performance and station readiness.

Designs are evolutionary and imperfect. Plant operators find that while some systems in a new plant are ineffective, some are superfluous and some quite wonderful. Reverse osmosis water treatment, rotary air compressors, and distributed controls are success stories. Ineffective designs are abandoned or redesigned. Redesign—if an economic choice—uses hard knocks gleaned from the previous inadequate designs. Users extend successful technology to other applications and facilities. Superfluous ones are left in standby or drift into obsolescence.

All designs begin with expectations, but operations alone prove them out. Not all designs or equipment provide the functionality sought. Lack of design alignment makes the post-design RCM backfit more challenging.

Component functions

Component function summarizes the output expected from components. Pumps pump but also contain liquid, provide ancillary service, or provide status or alarm. Many common functions are repetitive. Boundaries enclose fluids or power. Materials support structure. Instruments provide status. Component functions are usually reducible to 3–5 basic outputs. Valves isolate and contain. Safety valves isolate, contain steam, and relieve overpressure. Valve operators position valves to desired demand position. Air operators position valves with air as the prime mover in the operator assembly. Valve operators provide valve position status—how far open or closed positioned the valve operator is. Component functions and functional statements get more specific as component customization level increases.

For template-based RCM, component functions help to apply templates. Component functions and functional failures can quickly be assessed against system functional requirements to ascertain equipment safety, production, and cost implications quickly. By ranking functions and their failures by SOC criteria, the relationship to system functions is easily assessed. This ranking helps assign DFMs quickly, extending intervals where appropriate (*see* Fig. 2–21).

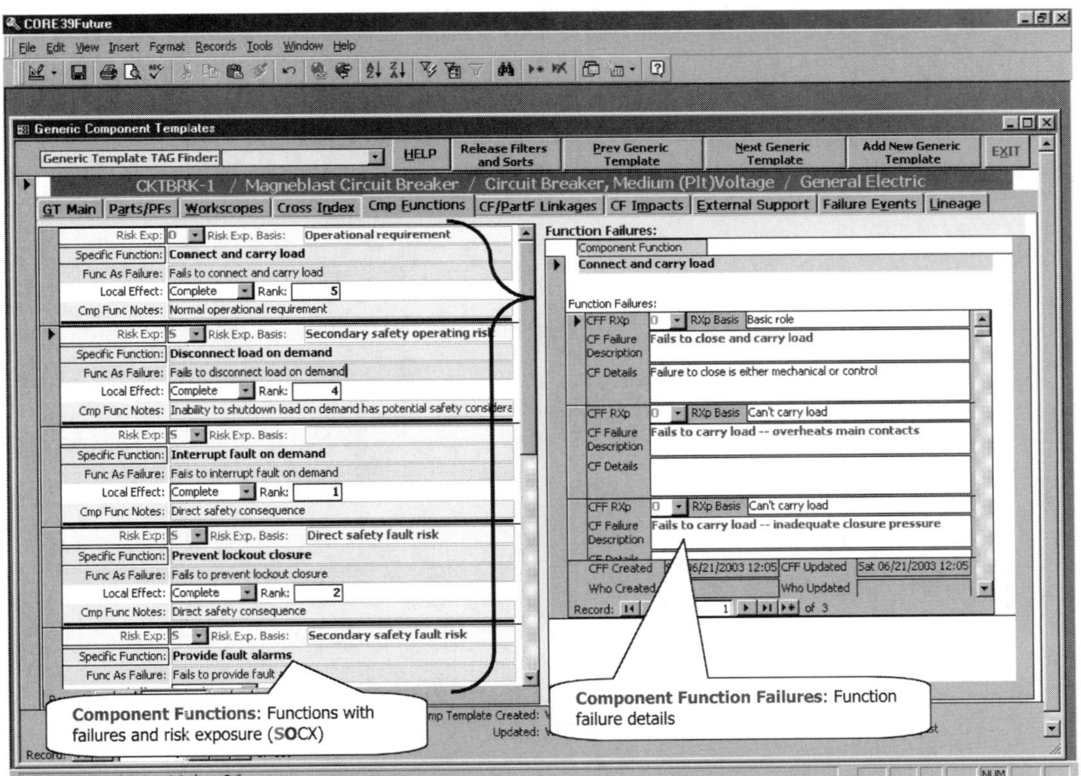

Fig. 2–21 Component Functions and Function Failures: Generic Template

Function alignment

In perfect design, equipment supports system functionality exactly. No equipment lacks clear functional purpose. Practical design lacks this perfect alignment. Equipment function use sometimes shifts because of operational learning and other factors. Sometimes equipment gets installed based on general operating experience and engineering thumb rules. All coal handling equipment gets dust suppression. All gas system isolation legs receive moisture traps. All air-water inter-coolers have drains. All controls receive redundant UPS power supplies.

Real plant equipment operates in ways designers don't always anticipate. These slight deviations between provided equipment functionality contrast with system requirements. RCM inevitably reveals little insights and operational changes that provide the basis to update system functional descriptions. (*see* Fig. 2–22) These in turn suggest design enhancements once operating goals are clear. Some organizations value the utility in better functional descriptions. Others won't.

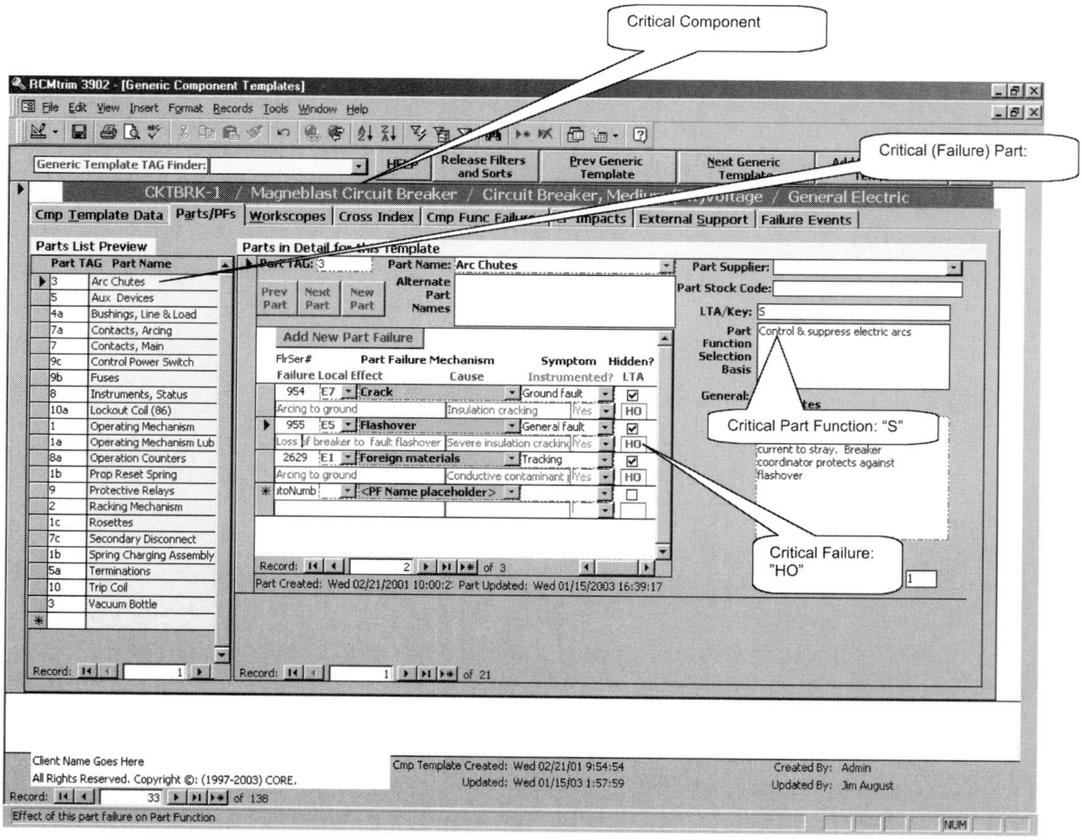

Fig. 2–22 Failures and Functional Effects

Some managers won't initiate changes for design or process improvements regardless of the benefits. Operating equipment misunderstanding can be contentious among operations, maintenance, and engineering departments. Where problem root causes and resolution options are not clear—and the tough cases aren't—RCM offers a way to establish a strategy and to find an optimum solution that best meets overall operational goals.

Equipment partitioning

Part breakdown structure ends at the largest, replaceable items, parts. Parts take analysis to the lowest tier, to the part failure. Failure analysis extends to the part level because parts are replacement units from warehouse stock. Workers replace or rework parts to restore equipment functionality. Part partitions relate parts (replaceable units) unambiguously to failures. Failures at the part level connect directly to the hardware that's responsible.

Components can be treated as *black boxes*, failing by losing function in multiple ways. Parts must be judged at the point of use by diagnostic assessment, for parts fail discretely, even in continuous failure. Component part breakdown structure extends the plant partition, delineating constituent equipment, parts, functionality, and how that functionality can be lost. The partition process allows parts—discrete, end-replacement items—to associate uniquely with failures for easier assessment, replacement, or rework.

While partitioning opens the black box, by identifying failure causes at the work performance level, it also facilitates decisions. By quantifying equipment, RCM aligns analysis to the physical installation itself. Viewed as a fault tree, part failure events initiate component failure. Identifying parts' failure modes allow replacement or rework. Unlike operators who need only recognize functional losses to identify failure, mechanics and technicians must deal with hardware. Maintenance mechanics and technicians must work *inside the box*.

Predeveloped, standardized component templates developed by parts, failure mechanisms, and their associated applicable and effective PM tasks make RCM development efficient. In large facilities, equipment design and installation symmetry can be exploited with templates streamlining PM analysis. Templates are useful on at least two levels: Equipment models out-of-the-book based upon *type* provide raw material for application. Type includes major type and subtype, like *pump* and *vertical*, or *operator* and *air*, which comprise *vertical pump* or *air operator* equipment descriptions. Approved standard generic templates may be applied based upon equipment installation context, design, usage, and symmetry to quickly model lower hierarchy plant equipment. Partitioning allows specific selection and application of generic templates. Selection chooses a template; application picks the DFM and adjusts template tasks by context for the equipment being modeled and analyzed.

Normal models

Normal models provide equipment reference cases, canons developed for identical equipment application. Normal models are based upon reference-developed templates. In a three-pump train in which one pump's plan identically applies to any one of three pumps, the second two pumps may simply refer to the first's scheduled maintenance program as the *normal model*. As a force majeure, normal models reuse analysis the same way that link-edit pointers reference other materials in MS Windows applications. The reference looks like the original.

Normal models require

- equipment dominant failure modes that are the same as those of the reference normal model equipment tag
- normal model and derived equipment user common operating and environmental contexts

Where operating symmetry is perfect, the two models are identical. Practically, perfect plant symmetry is idealized, but often closely approximated. Each normal model receives one *applied template* customized from a generic template to reflect the real application—installed plant hardware (*see* Fig. 2–23).

Primary secondary association

Primary associations further organize loop/skid secondary equipment, defining black box subsystems. Data associations complete the PM program. Reports and upload files present completed analysis for reviewers and CMMS/EAMS entry. Approved reports certify materials that eventually return to the CMMS/EAMS as finished PM program workscopes and tasks with basis and scope. Primary secondary equipment associations are a convenient way to combine multiple equipment tags that are functionally indistinguishable under one tag by dealing with the equipment as a single loop, skid, or assembly.

Primary-secondary associations have the opposite effect of new equipment tags to the MEL register. Combining tags with a primary-secondary association consolidates over-registered equipment. (Regulatory misinterpretation and zeal are contributors to over-tagging.) Tagging has little value below the component level. Associations cleanly ensure that every equipment tag in the register has been considered and appropriately dovetailed under another tag by association.

Design risk

Risk exposure classification considers experience, redundancy, design intent, failure probability, and available instrumentation. Designers seek robust designs, and experience counts. Starting from proven equipment designs reduces system integration risk. New system designs with new equipment make risk high. Pilot designs exhibit high risk.

A designer's ability to provide clear P&ID drawings, to intuitively understand design and performance objectives, and to apply practical experience makes design an easier but no less risky step. Experience and similarity are the two features that mitigate design risk.

Dilemma

In some context, virtually any component can become critical. "For lack of a nail…a kingdom was lost" is the clichéd fault-tree story carried to extreme. When failure barriers are breached, complex fault chains become possible. Following an

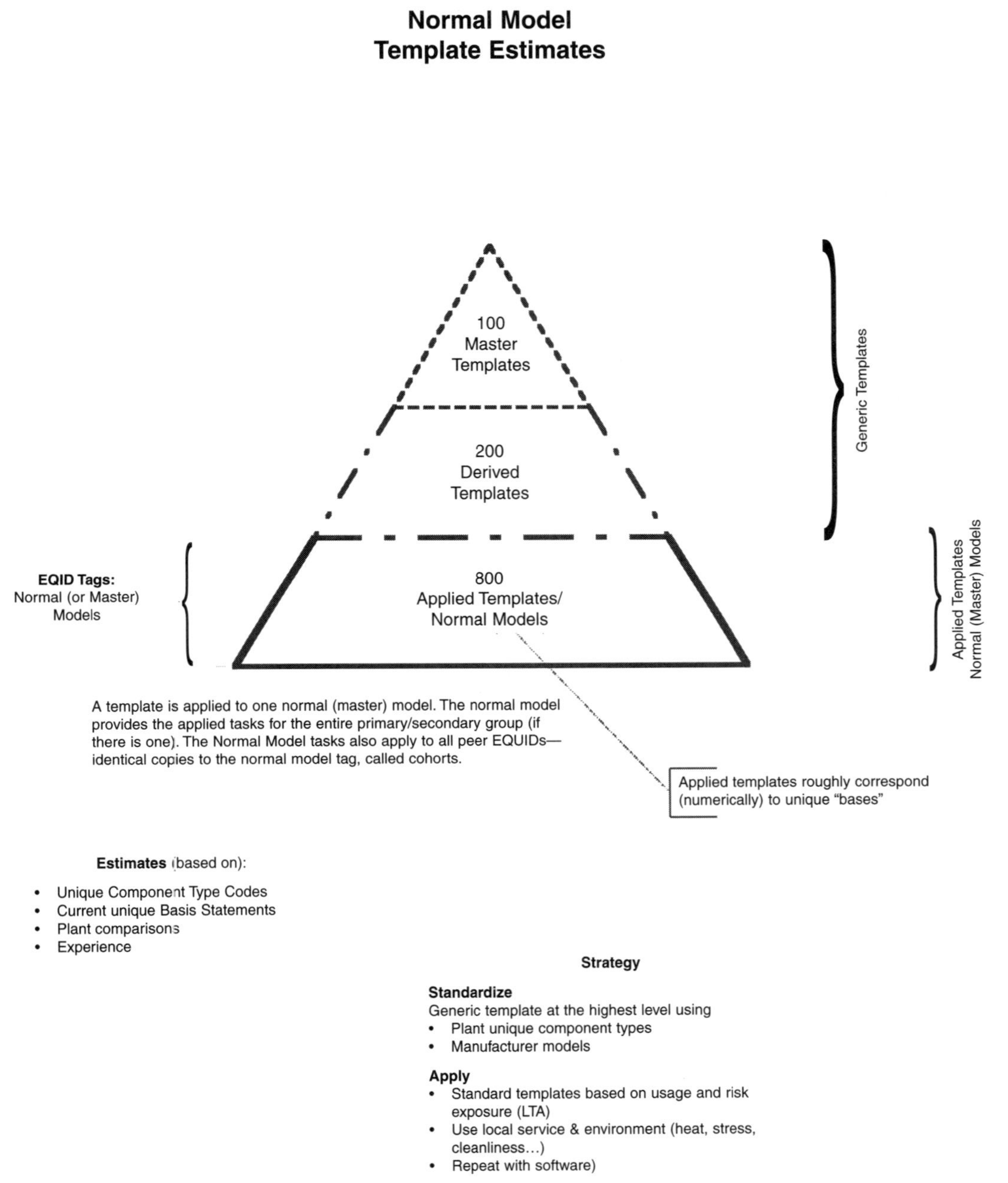

Fig. 2–23 Normal Model Template Estimates

event in hindsight, the barriers that would have been maintained were instead breached; backward-tracing event causes is a deterministic exercise. Simple, intended fault barriers are easy to identify in hindsight. Proactive barrier maintenance (e.g., breach avoidance) is more complex and provides one reason why RCM was developed. Direct failure restrictions impose significant constraints, demanding a substantial maintenance commitment. Performing on-condition maintenance and its derivative condition-directed maintenance when due is a primary failure barrier in modern plants using RCM-based maintenance strategies. Ironically, although these demanding commitments require discipline, they reduce overall work load substantially.

Selecting probable failures for redundant, conservative equipment designs is as hard as executing the associated selected PM tasks. For electronic instruments, failure to operate is often likely. Consequently for controls, redundant channels with voting schemes that select the correct inputs are critical. Where design redundancy comes into play, interfaces that select and isolate failed channels are a central to design.

Critical equipment design involves sophisticated features and tasks to assure that independence is maintained. Design features to assure independence occasionally fail. For example, a Boeing 737's hydraulic actuator with independent positioners in a common assembly has been implicated in a rare, sudden loss of control in the plane's rudder elevators.

Designs are never perfect. They improve with each operating year, but thousands of operating years are required to fully wring out a new design. Dominant failure modes identification is central to an RCM effort. Years of experience make engineers comfortable identifying probable dominant failure modes. In work, dominant failure mode selection is best performed by workgroups. This assures

- relatively certain occurrence
- practically described failure mechanism(s)
- integrated facility level experience
- decision conconcurrence

Environment and service strongly influence failures, both expressed and interpreted. The term *express* is based upon the term *expression* as used in genetic theory. (Genes may be recessive and expressed or unexpressed.) Unexpressed failures are inevitably present but never revealed over the equipment operating life.

Failure benefits when multiple eyes view components from a hands-on facility perspective. Developing generic lists of all possible failures assures no potential dominant failure goes unreviewed. The objective—leaving no potential failure excluded—leads to the generic template.

Listing all possible failures risks over-selects dominant failures unrepresentative of use—an unrealistic composite picture. Conservative aging interpretation based upon unrealistic life expectancy also over-prescribes work. Real equipment application scenarios have no average equipment! The solution rests upon risk and statistics. Most cost-based failures are evident in the work order history for units with several years or more of service. Operational- and safety-based dominant failure modes occur integrally but are often directly addressed by codes, standards, and insurance requirements. They are unlikely to be overlooked, obscured, or hidden based upon manufacturer guidance, standards like ASME's and the Institute of Electrical and Electronic Engineers' (IEEE) boiler/instrumentation programs, codes certified by law (ASME Boiler and Pressure Vessel (BPV) code), insurance guidelines, and other industry inputs. State, insurance, and industry oversight and are usually effective in identifying operational- and safety-based dominant failure modes. Identifying the risk exposure associated with each failure mode relates the owner operator's latitude to manage a failure mechanism.

Aging plants introduce new failure modes. Maintaining industry participation through professional organizations and operating groups helps plant engineers stay abreast of emerging issues. The ASME's reliability committee, Society of Maintenance and Reliability Professional (SMRP) reliability certification, EPRI, and nuclear industry groups are just a few of these professional organizations.

In reviewing critical components, analysts pick failure modes that have high probability of occurrence and SOC failure consequences. This is no simple task, for it leads back to design while using operating experience. Developing failure modes and mechanisms is part art, part science. Failure modes reduce failures to engineering descriptions. However, the best failure mode descriptions reflect hands-on descriptive insights. Failure discoverers—operators and crafts—provide these insights. Engineering failure descriptions are too arcane for practical use and provided at such a detailed level that often craft don't recognize them.

What are useful are descriptions that broadly classify group failures at a level suitable for the craft. To make failure descriptions useful, the craft should review failure descriptions and restate failures in their own terms. In this way *breaker phase-to-phase fault*, for example, becomes *breaker flashover*, and *pump pressure recovery loss* becomes *impeller blade erosion*. The latter descriptions are clear and relate part deterioration (failure mechanisms) to component function loss, e.g., the failure mode seen and experienced by operators. Craftspeople deal with hardware and parts; failure mechanism descriptions must express failure in part terminology to provide utility to them.

A failure mechanism combines failure mode and cause. Failure mechanism, root cause, and failure classification seek to simplify the so-called five causes of failure. Identifying failure root cause is a necessary condition to eliminate the failure. PM strives to control failures; elimination is a final design step. Redesign to remove failure is indeed a part of an RCM-based PM program option, but failure control with PM is usually more cost effective and, therefore, frequently selected as a practical maintenance step. One need not eliminate root causes to control failures.

By identifying part degradation that causes functional loss, a part can be replaced or reworked to control the failure. Identifying and replacing or reworking parts steps comprise the daily maintenance routine. One must recognize failure symptoms to identify failing parts and enable PM. In traditionally developed, craft-implemented maintenance, failure symptoms may not be understood at the decision-performer level. Unless failure symptoms can be clearly identified and recognized by the craft, on-condition scheduled maintenance is an elusive target. A significant portion of any RCM-based PM development effort is directed towards better failure symptom, effect, and cause identifications for improved failure recognition in the field.

Manual analysis

In performing RCM analysis manually for large equipment groups such as fluid-containing systems, one quickly discovers that similar functionality is repeated many times over within a common group of equipment. This knowledge suggests how to streamline processes to reflect the highly repetitious use of equipment in systems, while retaining the essence of RCM. Determining process streamlining methods requires reviewing the same process steps over and over. It demands patient analysis, and slowly analysts discover ways to reduce repetition. Developing a time reduction strategy leads to a strategy of developing and exploiting templates.

Failing to streamline and simplify, analysis results will look similar to the early generation studies of the 1980s—thousands of hardcopy pages per system, repetitiously performing the seven steps (or subsets) in exhausting detail. These studies are interesting to review from a historical perspective, but the reader is left scratching his head, asking, "How can I capture this system analysis succinctly without too much effort?" and "Must it really take 1500 hours of analysis to develop the RCM maintenance strategy for an average power plant system?"

Template depth and context

After performing manual analysis on a few systems, an analyst practically learns functional and hardware descriptions of common equipment well enough to compose a mental template. Capturing the analogous database template transforms the idealized template image for standard use, streamlined for repetitive application. This activity corresponds to reducing a system design concept to P&ID format; the concept is abstract, but the P&ID is concrete and supports more exacting use—like constructing a plant.

Template conceptually grows and deepens with use. The first template depictions come as two-dimensional lists of equipment with failure modes and characteristics. As the model grows, one finds that equipment requires more than two dimensions to model well. Placing a template in two-dimensional form quickly makes restrictions apparent. Templates have risk context that determines the most important failure modes in systems application. Real equipment in real systems performs real functions that determine a service context. Equipment in continuous service ages; equipment in stand-by service does not, or, at least, not in the same way. Finally, equipment has an environmental context; the location and conditions of use affect aging and performance. Rubber compounds in acid solutions suffer chemical attack. Hotter breakers wear out faster, and so forth.

Multidimensional template depth doesn't lend itself to two-dimensional representation easily. On one PM optimization project, the project manager selected two-dimensional spreadsheet templates to perform analysis with available Excel software. With five variables needed for application, dividing the spreadsheets into subsections developed additional dimensions. While initially simple, three months into the project, the practical outcome was unmaintainable data due to complex relationships between the dependent variables. Five analysts were mentally exhausted from visually interpreting and manually manipulating these complex spreadsheets. The human mind doesn't easily develop and maintain hundreds of n-dimensional relationships in two-dimensional form. (This problem was solved by the introduction of relational databases.) Virtually all analysts left the project within months of starting. The challenge to maintain data relationships suggests a better template format—a set of relationally connected tables!

Dominant failure modes selection

At the safety and operational levels, dominant failure modes are often clearly pre-identified. Code inspection and testing assure industry safety issues are addressed. Cost-based failure identification represents the target area of opportunity. Costs have strong operational, environmental, calibration, and alarm test consequences. Operators working with a full complement of instrumentation and fully functional alarms make better-informed operating decisions.

Suppliers and aftermarket test and vendor services providers define aging failure modes very well. Performance tests for pumps, fans, compressors, and heat exchangers help identify gradual performance losses based upon well-defined aging mechanisms. It's difficult to measure losses where two of three fans or pumps normally supply full-rated load. Instrumentation and control programs may require improvement where alarm limits drift or substantial resources are committed to equipment calibration on equipment for status-only operational use. Finding dominant failure mechanisms that exhibiting aging is relatively easy, and these failure mechanisms should rarely cause surprise nor pose engineering challenge. Drift, erosion, corrosion, and soft-part aging are all well defined at the engineering level. Most industrial plants today work with mature technology at the process level. There are few truly new failure mechanisms, and these are typically the subject of intense review at industry conferences.

Intervals

Once any particular equipment model (and its associated generic template) is selected, dominant failure modes are applied specifically to the component. The question becomes, "What adjustments are appropriate to that equipment's applied template model?" Making the transition from the idealized model—the equipment *generic template*, to real plant application—an installed piece of plant equipment requires the *normal model*. The normal model, which identifies the applied template, specifically uses generic template dominant failure modes and corresponding tasks in development. Actual installed equipment aging specifically reflects the normal model application. The application's context must be known to answer the question, "What are the right performance intervals?" This information allows one to develop the applied template—a template applied to a specific equipment context. Failure to adjust intervals for appropriate contextual aging adopts nominal manufacturer or vendor intervals, which leads to cantankerous scheduled maintenance programs in large facilities. These massive PM programs were the dinosaurs of the 1980s that couldn't be sustained economically or organizationally. They were just too aggressive and unnecessarily broad in scope.

Setting aging intervals determines the applied template aging context parameters, the natural age measure for the applied template, and adjustments that reflect the local scheduling processes available to assure the DFM tasks that prevent failure are performed successfully. This includes hard time to on-condition maintenance shift (and vice versa, if necessary). The task selection guidance found on the lower half of the summarized form of the MSG-3's task selection process should be considered for reference (*see* chapter 2, Fig. 2–8).

Task Selection

Traditionally, task selection started PM development. One located the O&M manuals, their maintenance sections, picked recommended tasks, and inserted them into CMMS/EAMS PM tables or WOs. In contrast, classic RCM PM task selection is the *last* step! Performing preliminary failure analysis puts task selection into proper context.

Consider a vehicle tire PM program. You would probably specify two tasks—checking tire air pressure and tread depth monthly and semiannually. Now consider the program with a non-rotated spare. Oops! You may have presumed the tire would be in continuous service, on the vehicle, aging. If it isn't, checking non-wearing spare tread wear has no value. Checking spare tire air is critical only before long trips or as a precaution at long six-month intervals, depending on risk. Context means everything!

Plant PM developers have identical context challenges—only much more complex. Few read the O&M manuals. Fewer have had a formal PM selection process training. Checking context provides the background against which vendor tasks can be intelligently examined, reviewed, and selected upon merit. Every vendor's O&Ms have one standard disclaimer, that "the user is cautioned to consider PM recommendations in light of their own service, environment, and experience." No equipment provider knows the user's context, and even AE's, designing a new plant, consider scheduled maintenance largely a guessing game! AEs build—but do not operate—plants! Operations accumulates the experience that fine-tune PM selections.

PM templates can help select PM tasks contextually. Selection in context means picking dominant failure mode tasks for failure modes that affect needed equipment functionality. Getting the *right* PM tasks requires knowing failure modes that are likely to express themselves, even in the absence of failure history. Identifying dominant failure modes and adjusting performance intervals appropriately remains the central task selection step. Interval adjustment requires failure mode examination in the operating context, as well.

Tasks

Equipment familiarity allows quick maintenance task consideration and selection. Lacking familiarity, extra care is needed, and more technical expert questioning is required. By bringing their equipment into commercial production, vendors do excellent jobs documenting their equipment's anticipated needs. Additional resources to identify DFMs and estimate aging and service intervals include industry literature available from sources such as EPRI, the U.S. Nuclear Regulatory Commission (NRC), the U.S. Occupational Health and Safety Administration (OSHA), the Institute of Nuclear Power Operations (INPO), the American Bearing Manufacturers Association (ABMA), Cooling Tower Institute (CTI), and the North American Electric Reliability Council (NERC).

Equipment models are based on equipment similarity and experience. Similarity may include parts, part failures, functions, and function failures. One may not understand centrifugal compressors, for example, but if they understand turbines (for dynamic blading), bearings, and gearing, they have the parts for a rotary compressor. Building concepts upon already known knowledge accelerates learning and development. Intuitively, since centrifugal compressors and turbines have many of the same parts—dynamic blading, bearings, and gearing—it is reasonable to presume they act and fail in similar ways. Applying existing template developed parts, failure modes, and other component details helps quickly build new templates by similarity. At the part failure mechanism level, there are many ways to connect the pieces. Fundamental failure mechanics remain the same. After a few years of reliability analysis, encountering a really new equipment design becomes a rare treat! Once drafted though, the developer should always seek out those with hands-on experience to validate the rough template. In most cases, some interesting twists or unusual insights unique to the design will be added.

High-speed compressor blade erosion occurs very quickly. Scaling and male-female screw compressor elements introduce cavitation erosion unique to each machine. Fortunately, manufacturers' designers know the peculiarities of their machines very well. They typically know expected part lifetimes, part limiting-life wear-out mechanisms, and a host of other useful information that fails to make the O&M manual cut. Establishing maintenance plans for their equipment before plant startup is a proactive and insightful effort. Insightful details are a part of the proprietary knowledge base provided by an original equipment manufacturer (OEM). Capturing these details while developing a maintenance strategy adds value. Ideally, this happens during startup before a plant comes online. Practically, it should continue over plant life.

Once risk exposure has identified critical equipment, task selection occurs. Some general rules apply. Do simple things before complex ones; perform simple checks before overhauls. The logical selection progression developed in MSG-3 provides a simple path: In order, address failure modes, select one task for each dominant failure mode that is applicable, and effectively addresses failure (*see* Fig. 2–24).

For instance,

- simple lubrication and service
- operational checks and tests
- condition monitoring (on-condition) checks
- failure finding tests
- rework or replace
- redesign

In all cases except safety, it is correct to seek a single applicable and effective task that addresses each equipment part failure mode. *Applicable* simply means works; *effective* means *works cost effectively* and *efficiently*. Works requires statistical analysis to determine the times the job is performed is successfully. Practically, some engineering inference and interpolation is required. That *works* means

- accepted in the industry
- not developmental technology
- skilled, trained technicians working with quality tools achieve the same result
- quality time and support conditions are available
- field conditions (people, environment, tools, skills, etc.) prevail

Some technologies have only recently crossed the effectiveness threshold. Thermography and partial discharge (PD) analysis are two. Vibration monitoring met the requirement in the 1960s. Other marginally effective technology is improving daily. Just as radiography requires skill, so do other technologies. However, just because one can service equipment doesn't mean that person is proficient in overall maintenance task performance, equipment diagnostics, or PM development theory. Proficiency is assured by training and qualification testing, not title.

Thermography provides a case in point. Effective thermography use requires application knowledge like mass flows, power flow, and the energy principles. Although a technician can quickly be trained to carry a thermography camera around to take pictures, it takes an understanding of internal energy flows and likely problems to interpret them. Like the NDE technician shooting welds, taking the picture is one thing, interpreting the result quite another.

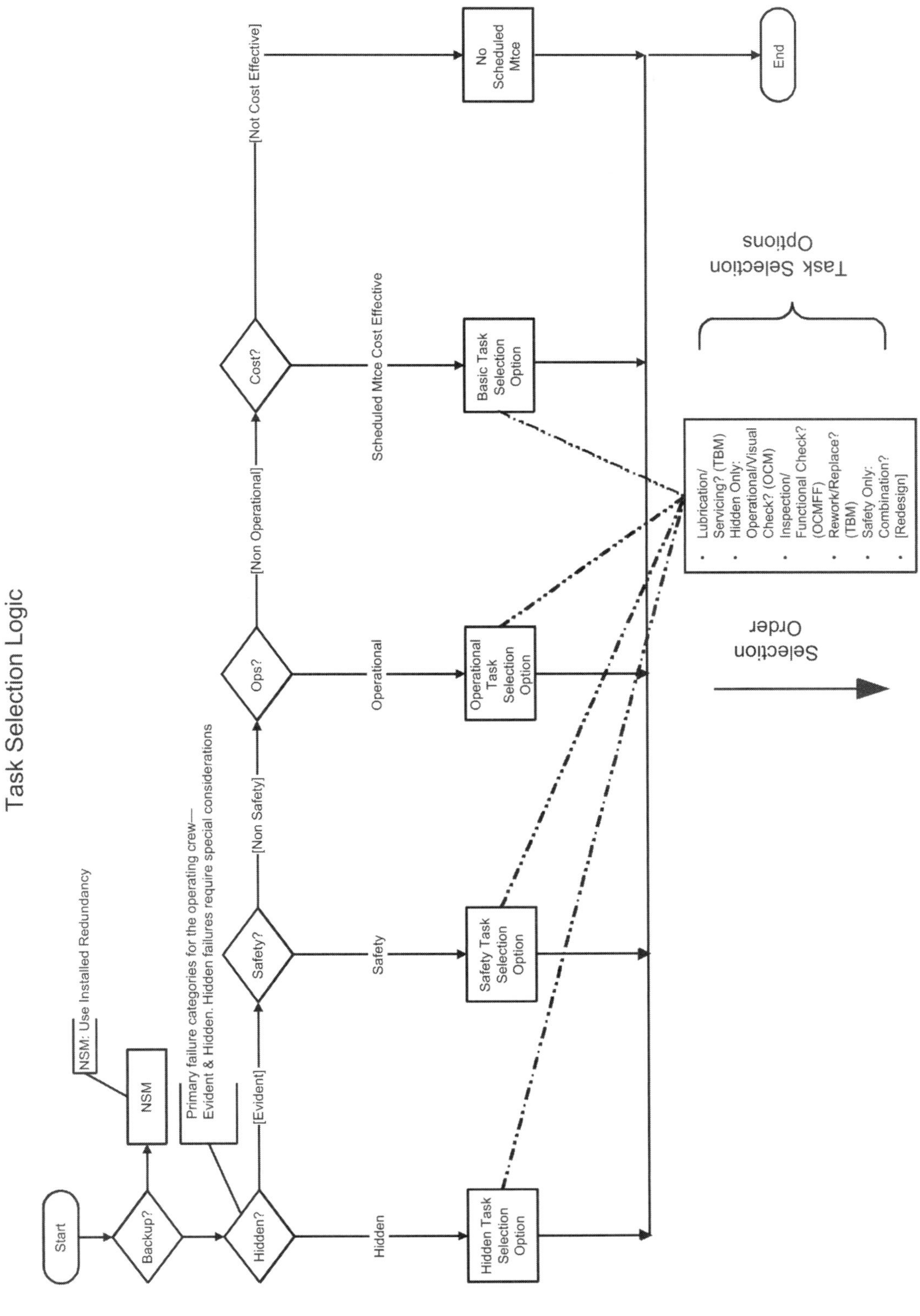

Fig. 2–24 MSG-3 Detailed Task Selection Logic

Engineering Failure Mechanism		Technology	Characteristic Measured	Detects
Elastomer	A	1. Durometer 2. Visual (color texture)	1. Rubber Elasticity 2. Cracking/Chemical Attack	1. Cross-linking deterioration 2. Chemical deterioration
Balance	B	1. Vibration Analysis 2. Noise (Audible/Ultrasonic)	1. Acceleration 2. Rubs 3. Alignment	1. Imbalance 2. Contact 3. Misalignment 4. Looseness
Corrosion	C	1. Visual 2. Thickness 3. Composition Change	1. Composition & cracking 2. Remaining material 3. Composition	1. Chemical corrosion attack (general /local); oxidation; fatigue 2. Wall thickness/fouling 3. Decomposition
Deposit	D	1. Visual 2. Thickness 3. Composition Change	1. Deposition 2. Additional material 3. Composition	1. Thickness material changes 2. Loss/change in heat/flow capability 3. Loss/change in heat/flow c capability
Electric	E	1. Resistance 2. Turns ratio 3. Doble (inductance) 4. Magnetic 5. Current signature 6. Hi-pot resistance (AC&DC) 7. Voltage trace (visigraph) 8. Reflectance (TDR time domain reflectometry) 9. Partial Discharge 10. Combinations	1. Resistance Drift/Open 2. Shorted turns(transformers/windings) 3. Dielectric insulation changes 4. Ferritic core changes 5. Dynamic current inducements 6. Insulation changes/moisture 7. Dynamic operations behavior 8. Dielectric weakness along a path 9. PD events; winding PD buildup 10. Synergistic characteristics that signal specific failure modes	1. Resistance 2. Turns ratio 3. Doble (inductance) 4. Magnetic 5. Current signature 6. Hi-pot resistance (AC&DC) 7. Voltage trace (visigraph) 8. Reflectance (TDR time domain reflectometry) 6. Combinations
Filtration	F	1. Differential Pressure 2. Through flow 3. Deposit activity 4. Other	1. Differential Pressure 2. Through flow 3. Deposit activity 4. Other	1. Differential Pressure 2. Through flow 3. Deposit activity 4. Other
Lubrication	L	1. Viscosity 2. TAN (acid number) 3. Particle counts 4. Disolved metal species 5. Moisture content (dissolved) 6. Moisture suspended 7. IR Spectrometry	1. Basic property to stick and flow 2. deterioration 3. Stability of lubricated surfaces; metal to metal contacts 4. Metal removal characteristics; catalysis reaction with particles. 5. Moisture contamination 6. Moisture suspension emulsification 7. Chemical package depletion	1. Fundamental ability to lubricate 2. Remaining life as lubricant 3. Metal contact among lubricated parts; rubs, hammering and misalignment. Hard starts 4. Chemical package depletion 5. Moisture content (dissolved) 6. Moisture suspended 7. Aging of lubricant

Fig. 2–25 Technology Comparison: Engineering Failure Causes and Diagnostic Options

Metal	M	1. Alloy content 2. Weld radiography 3. NDE Magnetic Particle 4. NDE Dye Penetrant 5. NDE Ultrasonic 6. NDE Replication 7. IR Spectrometry	1. Alloy composition 2. Weld cracks, inclusion, . . . 3. Fatigue cracks starting at surface 4. NDE Dye Penetrant 5. Deposits, loss of material, two layer oxidation growth 6. NDE Replication 7. IR Spectrometry	1. Alloy content 2. Weld radiography 3. NDE Magnetic Particle 4. NDE Dye Penetrant 5. NDE Ultrasonic 6. NDE Replication 7. IR Spectrometry
Environment	N	1. Temperature 2. Pressure 3. Moisture 4. Other chemical (atomic H, ozone,...)	Fundamental physical properties	Environmental shifts
Control	O	1. Range 2. Hysteresis deadband 3. Dynamic change response	1. Setpoint drift 2. Wear, component drift 3. Inherent response	1. Range 2. Hysteresis deadband 3. Dynamic change response
Friction	R	1. Ultrasonics 2. Infrared Thermography 3. Audio sonic	1. Rolling element friction 2. Heat from friction 3. Friction audio rubs and contact pitch	1. Untrasonics 2. Infrared Thermography 3. Audio sonic
Structural	S	1. Dimensional measurement 2. Crack growth 3. Stability (Survey) 4. Visual	1. Corrosion material loss/settling 2. Load shifts or growth 3. Settling 4. Settling, environmental, chemical attack	1. Settling/loss of load capacity 2. Load capacity loss 3. Support foundation shifting 4. Environmental or site change
Temperature	T	1. Embedded TC, RTD direct 2. IR indirect	1. Temperature rises from inadequate heat removal/excess generation 2. Heating	1. Excessive loading, loss of cooling, aging trends 2. Energy flows or flow shifts that displays comparatively uneven in visual picture format
Fabrics/ Composites	W	1. Visual 2. Spectometry 3. Surface reflectometry	1. Cracking decomposition 2. Chemical changes 3. Corrosion/chemical decompositon	1. Cracking decomposition 2. Chemical changes 3. Corrosion/chemical decompositon
Consumed	X	1. Thickness 2. Level/Pressure 3. Color	1. Remaining material /deposit accumulation 2. Remaining fluid amount 3. Contamination/Moisture	1. Lost material / accumulated material 2. Amount left 3. Dessicant depletion (moisture)
Dynamic	Y	1. Ultrasonic, Venturi, Vortex Flow Measurement 2. Load Capacity 3. Response	1. Flow 2. Load (Often composite calculation) 3. Response	1. Wall roughness or fouling flow loss 2. Heat transfer load; other load 3. Control loss, control element component deterioration & aging (like control valves)
Unknown	U	1. Radiological 2. Nuclear	1. Radiological 2. Nuclear	1. Contamination, dispersion, corrosion activation 2. Nuclear depletion, poisoning

Fig. 2–25 (continued)

Sometimes a proven technology is already recommended. Vendors usually know aftermarket technologies used on their equipment and can suggest options. Some even provide predictive technology or services. Consulting engineers are also reliable, technically competent resources. Sometimes, failure-cause study suggests an obvious predictive technique. Dimensional checks apply in wear applications. Knowing degradation mechanisms that are occurring and the technology available suggests where to look for diagnostic tools. Operators and mechanic/technicians who work with or on the equipment also have insights about failure symptoms such as noises, smells, and visual cues. The five sensory inputs are often extremely insightful about where to look and, indeed, whether failure is even evident. Where hidden, instrumentation and sensing may be necessary (*see* Fig. 2–25). This quickly becomes heavy design stuff!

Engineers who develop equipment thumb rules carry equipment templates in their heads for the equipment they know. Using this mental image model provides a foundation to build upon. By extending templates, a reliability analyst with plant experience can build a custom set of common plant equipment templates from scratch over a couple of years with minor, incidental time input. Developing an exhaustive, comprehensive template set for a complete facility should take no more than two years. After several years, the reliability program effort should shift to address ongoing strategy retention for previously analyzed equipment.

With a predeveloped template in mind, the engineer or analyst can select dominant failure modes from a part breakdown structure. Failures either apply based on experience or must be highly probable over the facility lifetime based upon similarity analysis and expert opinion. Opinion requires some interpretation and works best when a team provides it. This extends available installed hardware operating contexts to cover likely failure modes when data isn't available.

Task intervals

Selecting task intervals is part engineering and part art. (A competent relay engineer once said that relay settings were mainly art, and he was the only engineer willing to paint the setpoint picture!) The same applies to setting any PM task intervals. Many engineers are extremely reluctant to make any personal assessment and put their name on it. To do so professionally means they are confident they know their equipment! Most truly critical safety applications have code standards that remove discretion. Operational production risks can be monitored with judicious benchmarking to reveal when intervals are aggressive. For cost-based work, the aggressive extension of intervals towards failure limits is obligatory, particularly where excessive PMs discredit the overall scheduled maintenance process. This is often the case!

We have a Catch-22 situation: Workers skip PM knowing full well that scheduled intervals are too conservative, but extending PMs to proper intervals without resetting mental attitudes risks blowing by the schedule limits using the old paradigm. Again, identifying SOC criteria, when cost is the driving factor, removes that risk of revising or removing a safety-based task. Workers can receive the same task criticality flags on WOs. Statistically, SOC stratifies criticality. Roughly 50% of PM applications are based on cost, 30% on operations, and the balance (roughly 15–20%) on safety. This stratification is the reason why SOC criteria are so useful. They generate a pyramidal hierarchy of tasks that supports prioritization by risk!

Task interval selection begins knowing the risk category. Safety critical equipment almost invariably has specified limits based upon codes, insurance carriers, or regulatory bodies. Licenses specify how often key safety inspections must be conducted. Following regulated guidance is easy. Identifying other critical risks—operational or cost—are not once the burden of safety is lifted. Operational limits can be benchmarked or reflected against operating history or insurance standards. Comparisons to the same equipment failure modes in safety applications may be useful. EPRI and vendor literature should be consulted.

Once these sources are exhausted for those risks that lend themselves to data collection, consider preparing histograms of expressed equipment failures. Histograms of failures contrasted statistically tell the story. Whenever operating losses or production costs are involved, histograms of these losses should be prepared. Depending on criticality type of the failure mechanism/PM task in question, conservatism based upon risk sets the PF interval (*see* Fig. 2–26).

Selecting PF intervals depends on risk tolerance. Again, where condition assessment technology is very well developed (like vibration monitoring or oil sampling), shorter intervals may be appropriate. When uncertainty abounds, longer intervals

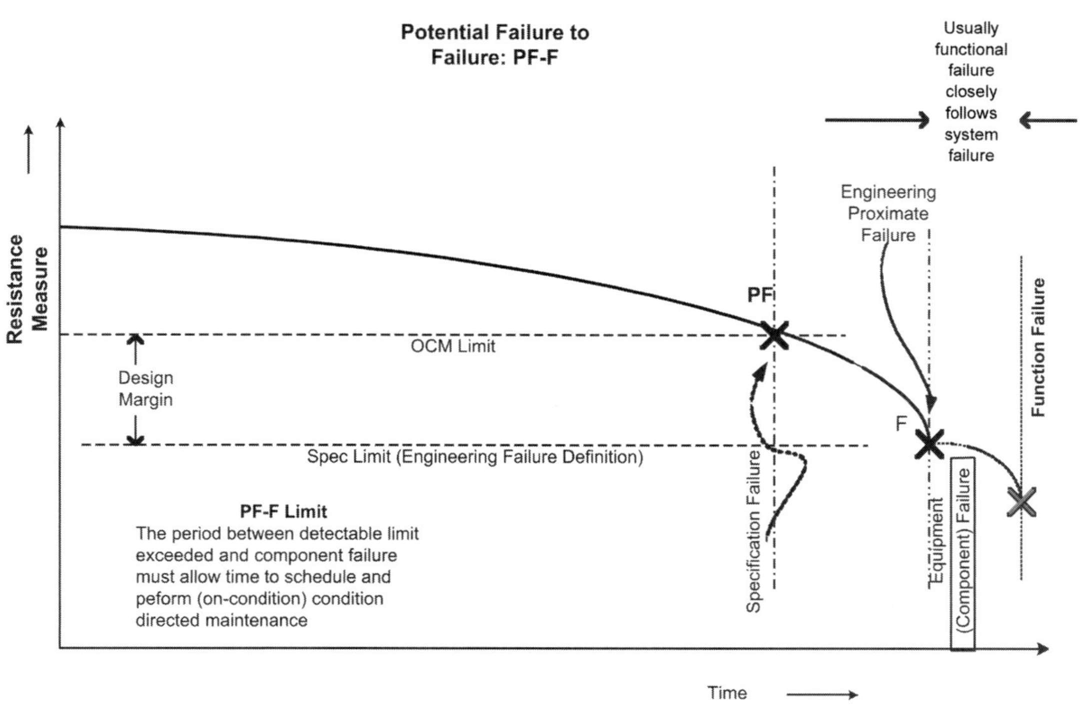

Fig. 2–26 PF-F Theoretical Graphs: Window Between Failure Indication and Failure

should be specified. Finally, the organization response time necessary for performing corrective maintenance should not be overlooked or presumed. Some plants need long lead times to perform indicated work. That needs consideration in developing PF intervals for condition-directed maintenance.

Workscope application

Workscopes need to accommodate as many tasks as reasonably fit into a given work interval. Practically, any equipment work period is a special trip or outage period, at least at the equipment level. This unique opportunity to view the equipment—test it, open it, work on it—needs to be treated with economy. Trips must be reduced, tagouts minimized, tests consolidated in order to maintain operational control and economy of performance. The workscope provides the conscious step of assembling tasks into larger blocks for efficient performance (*see* Fig. 2–9).

Blocking tasks for workscopes trades perfect, ideal task timing for organization and economy. Grouping work into workscopes sacrifices intervals a bit to achieve performable work blocks. The need for developing workscopes was discovered more than 30 years ago in aviation RCM. Any time the cost of taking equipment out of service is high, which is invariably the case in industry, the work must be grouped for maximum advantage. (*see* Fig. 2–27) The workscope provides a convenient scope of work (in project management phraseology) that defines what has to be done to complete the scope. In PM, getting the worker mindset focused toward formally completing a specific workscope avoids the *dry-lab* syndrome that has lead to some monumental foobars in the past.

Comparisons

RCM-streamlined reality-centered maintenance (SRCM) (traditional)

Studies developed by the EPRI and others in the early 1990s, sought to gain acceptance and support for SRCM. They cited minor differences in outcomes between the classical and streamlined RCM approaches. Proponents of traditional RCM have questioned their validity and conclusions. The challenge is to make simpler processes do more. No doubt the market will decide which processes offer the greatest value. They will be the processes that capture the benefits of classical RCM in abbreviated performance. While they may not follow the exact format of EPRI-approved or vendor-specified RCM (or SRCM), they will capture elements of traditional RCM in elegant, brief packages.

By knowing the expected equipment service periods, an RCM (or SRCM) plan tries to discover on-condition monitoring and service prediction methods that extend intervals out to the known lifetime. This requires knowing equipment nominal life

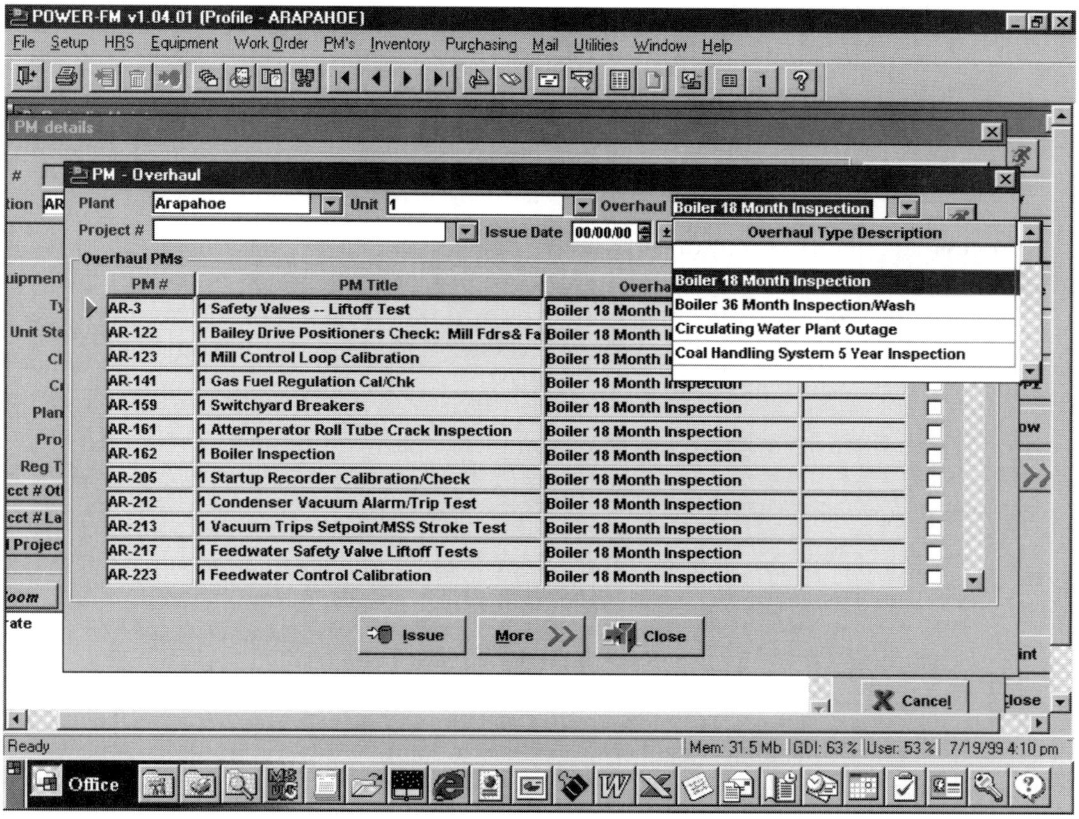

Fig. 2–27 Assembly of Workscope Into Larger Outage Plan

based upon the life-limiting characteristic and finding a suitable on-condition predictive technique that offers a PF-F interval that will allow servicing prior to failure. RCM or SRCM offer comparable results.

Streamlining: pro and cons

Streamlining requires profound process understanding, identifying the most time-consuming process steps, and focusing improvements on reducing step time contributors. Time-intensive steps must be evaluated for these opportunities: accelerated performance, trimmed requirements, replacement, or even outright elimination. Using a template approach, the most time-sensitive RCM processes activities are the following:

- system study to identify function statements, classify, and justify rough equipment risk exposure classification
- equipment review for further critical SOC risk exposure classification, general and specific failure modes)

- dominant failure selection (e.g., template failure mode task selection and application for failure mode task selection) using equipment risk exposure and service context based upon specific failure modes
- expert reviews: system and maintenance performers with direct responsibility for performance, maintenance decisions, and resources

Process acceleration requires reducing steps, reducing performance time, and improving productivity so that analysts achieve more results in less time. Ideally, those results constantly improve in quality compared to their precursors. Streamlined RCM processes and their critical review by expert RCM analysts and engineers stress common themes:

- Streamlined RCM can overlook complete analysis findings.
- SRCM can be a crutch for superficial analysis.
- Traditional RCM, manually developed, is too expensive for most plants to perform and virtually impossible to sustain a living maintenance process.
- Streamlined approaches are effective in most contexts.

Like an operational amplifier—that little miracle of early circuit board electronics—a little feedback can compromise a lot of gain yet remain a powerful reproduction tool. Modest amplification sacrificed gain for faithful signal accuracy. Well-developed streamlined and traditional legacy RCM processes fit this analogy. SRCM with feedback can produce some phenomenal results in a living program. This is a very good tradeoff, indeed!

3 Generic Templates

Component Template Strategy

Large maintenance programs require force multipliers to systematically develop consistent maintenance plans. Equipment templates provide one such force multiplier. Scheduled maintenance component template strategy is recognized by standards such as INPO AP-913 Equipment Reliability Process Description to assure maintenance consistency and effectiveness while developing consistent, cost-effective maintenance plans. Templates trade installed equipment variability and complexity for a predeveloped solution. Template application must appropriately consider these limitations.

Spreadsheet component templates have been used in RCM-based PM development for nearly 20 years. These two-dimensional flat files cannot develop multidimensional relational tables simply. Tables capture rich normalized data relationships in a database. Relational tables more easily portray the RCM depth needed to model different aging, risk, design, and performance factors. Relationships can interface to CMMS/EAMS' relational scheduled maintenance tables to automatically load PM program plans. These database middleware speed the final uploading steps into the CMMS (*see* Fig. 3–1).

Template means pattern. Templates model component maintenance plans. They provide common, prepared source patterns. Template utility stems from common design and widespread use potential. A vehicle maintenance template could apply to most highway vehicles. Common vehicles with oil sumps, windshields, and tires could be covered. Once a new vehicle variant had been discovered, a new template would be sought. Using a template effectively requires understanding template application limits. As new equipment extensions are found, the template should be extended to cover them.

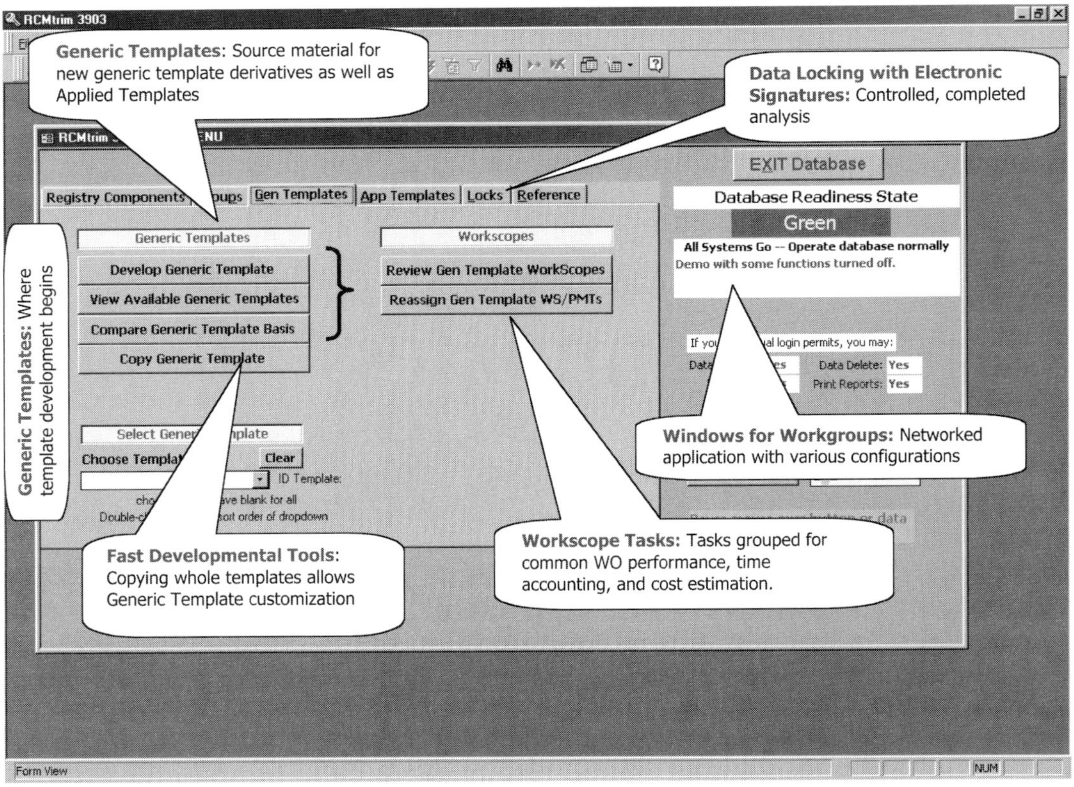

Fig. 3–1 Generic Template

Equipment design

Templates summarize similar equipment maintenance programs. Template success depends on appropriate design classes. More detail lessens template generality. A generic template balances generality and detail so that end-use influences development. The template must model the equipment so that a worker recognizes component design for work scheduling, tagout, and control purposes.

Templates provide unambiguous equipment description of component parts, failure modes (as observed by an operator), and PM tasks listing dominant failure modes. Templates impose several challenges. The parts breakdown schedule must be general and complete. Dominant failure modes must be organized around parts that can be tested, inspected, replaced, or restored. WO task intervals must independently group specific PM tasks to facilitate work planning and replanning—a common maintenance requirement. Finally, PM tasks themselves must be nominal at the generic level but should adjust with the part-failure-task record string to the applied context.

For example, an identical pump could be in a 1300 MWe generating plant, a smaller 350 MWe unit, or a chemical refinery. Each context would have unique risk exposure, dominant failure modes, PM tasks, performance intervals, and basis. All should be derivable from the same base table's template.

Starting Point

A blank or generic template provides a starting point for using and applying templates. Generic templates provide appropriate tasks to select and apply based on specific component context. *Specific context* is a specific piece of equipment in a specific field application. A pump book can describe pumps generally, but real pumps have context, i.e., a unique history that reflects risk, use, application, culture, and plant installation. Template application requires understanding the general model, plant context, and relevant failure modes (e.g., dominant failures) to apply in the actual case. Application must capture relevant hardware genealogy, usage, and stresses for the end user.

Defined arbitrarily, the source template is generic. It is broad, generally applicable, but abstract. It facilitates application without specific plant context. It conveys enough detail to model the actual plant installation quickly, providing cues allowing task selection and interval adjustment.

Finished work

An ideal template represents finished analytical maintenance raw material. It should be a perfect model—a model case that summarizes general equipment representation. Equipment can be partitioned into parts, some of which matter, most don't. A part breakdown structure for even modest-sized machines often lists thousands of replaceable parts! In some context, almost any part can matter, but in common usage, most don't—at least in the sense that they are likely to fail.

Problems in western states, like switchgear dust, aren't as pronounced in eastern states. Acid mine runoff problems, like carbon steel pump impeller and casing waste, aren't seen with more prevalent, non-aggressive raw water makeup. No matter how generic a template, users always need one more failure mode. The scheme must be extremely flexible for modification and adjustment of the template's parts, failures, tasks, intervals and their basis.

Generic templates capture, for a given class of equipment, parts failure modes that are reasonably likely to occur that can be prevented by scheduled maintenance. Rote detail accuracy should balance with general part failure descriptions that allow template material reuse. Template engineering is an art. Useful templates are familiar to the equipment users, the skilled craft that service the equipment. They must reduce technical jargon to meaningful terms recognizable by common users. That's a tall order!

Clear templates provide benefits later and take effort to develop. Clear, unambiguous dominant failure descriptions, failure symptoms, inspections, test acceptance criteria, and appropriate tasks compound benefits many times over. These criteria provide exact symptoms, limits, and task descriptions that make PM task performance actionable in the field. These raw materials assure that all scheduled maintenance in the field meets the same high standard.

Every maintenance facility has a performable task capability repertoire. Different technology (or similar technology from different vendors) may be available preferentially to address certain failure modes. Electrical insulation testing can be accomplished several ways, depending on available technology. *Doble* dielectric resistance testing, *turns ratio* tests, and *hi-pot* (high-potential) tests are just a few examples. Test technology terms reflect commercial development proprietary processes and trademarks.

The use of microprocessors to perform batteries of tests under common commercial application represents another twist on available test capabilities. Facilities need to be able to identify suitable equivalents or alternative test methods to the standard template-proscribed tests based upon tools and skills they have available. Application should allow substitutions of one technology for another, provided they both meet applicable/effective test criteria. It should also allow for the selection of hard-time tasks in instances where on-condition ones won't do. This flexibility in addressing how part failures are handled for a specific application greatly improves field program flexibility where conditions aren't always ideal, predictable, or even understandable at the time the maintenance strategy is first crafted. Accessibility to equipment, people, and techniques are limited in boilers, contamination areas, and other restricted-entry locations that reflect the real word, not lab conditions.

The generic template should carry the canon test (method for dealing with a failure), following the selection order of MSG-3 (*see* Fig. 3–2). Complex lube-oil sampling and monitoring are specified for large-volume lube-oil reservoirs. Where simple lubricant change-out is cost-effective and preferred in situations such as small bearing sumps, this should be specified, rather than lube-oil sampling and monitoring specified for large-volume reservoirs. Where hi-pot testing is preferred, it should be noted. Application flexibility should allow point-of-use changes to correct for actual equipment use, schedules, and needs.

Practical template evolution

Templates originated with manufacturer work descriptions and were formulated into text work orders in early mainframe CMMS/EAMS systems around 1980. A one-paragraph breaker inspection provided by General Electric for a MagnaBlast breaker would be uniformly used for Westinghouse, ITT, and other air-blast breakers of the same 4kV class. Field workers interpreted these inexact models where necessary. Despite obvious drawbacks, experienced electricians could interpret manufacturers' terms; interpolate settings and limits; and adjust specific workscopes, tasks, and intervals on the fly. It wasn't a bad start, and standardized workscopes gradually evolved. Though standardization was laudable, an early drawback included general or incorrect values and references that field workers or their foremen had to correct.

A more exacting approach began with standard template tailoring results for another application (the Westinghouse DB50 breakers, for instance). Compared to the previous generic case, more specific details and refined tasks customized generic plans

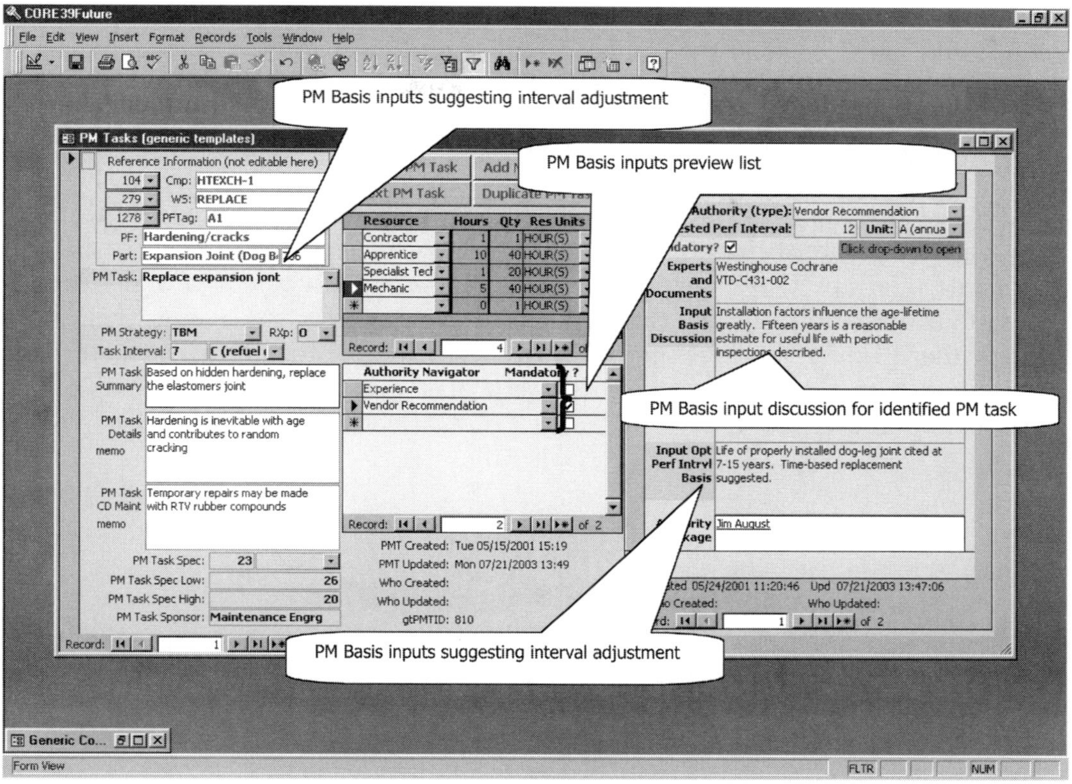

Fig. 3–2 Generic Template Logic

to an actual installation—an application-specific effort. The process came to be described as *application*. Many plants eventually generated detailed sets of exact-task workscopes in many equipment-associated procedures. Specifically tailored PM application template use increased. End users varied WO workscopes and tasks for applied templates provided to the craft. While large plants with abundant planning resources could develop detailed PMs, others with fewer staff or smaller budgets couldn't. Those that couldn't, relied on the skill of the craft to interpret WO's vendor manual details to supplement or adjust specified applied PM WO workscopes.

Even where plants had identical equipment and practiced standardization, installation provided another challenge. Those 4kV breakers could be customized and used for either load or feeder applications. Feeder application is more important, as feeder loss compromises the whole load center, but breaker loss only risks loosing the powered load. In some applications, loads trip more often. Ash water pumps in coal plants easily clog; their breakers trip out first, starting deadheaded pumps. Some loads power motors. Others feed transformer loads to lower-voltage power supplies. A vendor's generic guidance provides nominal, worst-case, conservative equipment applications. Using these directions, overly frequent service intervals loaded shops with servicing regardless of equipment aging or loss-risk. Regulators were frustrated that plants could have so many programs and still not "get it right," introducing more regulatory errors through maintenance itself.

Spreadsheets like Lotus 123 and Excel offered improved capabilities. Template development moved into two dimensions, but problems remained. Flat-file spreadsheet template formats limited dimensions. Spreadsheets that forced three or more dimensions into their two-dimensional sheets with defined ranges were difficult to read and interpret. One example of a cumbersome industry reference application is found in PM basis manuals and their associated database. A cumbersome application requires more time. These complex spreadsheets were difficult to use in project applications with multiple users. They were also difficult to control, subject to interpretation at every turn.

Building Generic Templates

Resources

Resources needed to construct complete templates from the foundation up vary considerably. Two skills are required to build successful templates: equipment knowledge and RCM proficiency. Constructing useful templates requires enough understanding of equipment in order to focus on parts and dominant failure modes of interest—those that generate component failures. The developer needs a fair amount of depth with the equipment, or the ability to research and interview to extract that knowledge from others. The ability to summarize and convey component and part thoughts, especially failure descriptions, in the workers' vocabulary is also useful. Using their language makes failure recognition easier in actual plant work. Once the template is drafted, it's important to review it with the intended users—maintenance planners, workers, and experts to assure the right thoughts, considerations, and descriptions have been captured.

Steps

Building a generic template begins by defining the component of interest, including its scope, functions, and part partition. It can be difficult to identify appropriate boundaries to limit the scope of a template, particularly when the equipment is built up on skids. Engineers are usually prompted to develop a generic template by a specific plant need. Therefore the effort often starts with a specific component in mind as the intended *target* component with the idea of broadening the scope of the analysis to be more general for similar plant equipment. This approach helps get the process going, but it also biases the template design towards one piece of hardware. Acknowledging this and moving forward is the easiest way to maintain PM momentum in projects. Plants can easily revise and customize completed work later, especially with database application software.

Creating the equipment parts partition can be as simple as pulling a parts list off the vendor manual parts list. Fortunately, most parts never fail or are replaced prior to failure by an overhaul or other task performance. The parts of interest are those that have dominant failure mechanisms that can affect component performance. Very often, these parts are identifiable from vendor technical literature because they are the parts tagged to be replaced during overhauls or other maintenance activity.

Vendor O&M manuals are a primary development resource and should always be consulted if available when constructing a template. They also are traditionally structured and rarely identify the failure modes that cause maintenance explicitly; failure modes must be inferred. They must be read with care, for occasionally key requirements are tucked onto one-liners or placed in operating information sections. Normally, the maintenance PM subsection has the most value in developing the planned maintenance program. Troubleshooting sections are also extremely useful because they identify anticipated failure modes and may even shed light on the operating and maintenance philosophy surrounding some failure. Emphasis on troubleshooting in lieu of other requirements suggests the vendor anticipates controlling random failures as the main maintenance requirement.

An effective way to review vendor manuals for complete technical recommendations is to scan the manual from front to back, placing emphasis on the O&M section, and highlight everything that addresses part replacement, time-based maintenance, and operating activity. Of course, where the manuals are controlled documents maintained by a library, the more considerate approach is to obtain a package of the small Post-It Notes® and tab all pages that address the same O&M section monitoring or time based maintenance requirements. A schematic or plan diagram and even a picture of the component is also helpful for reference. After the manual is tabbed, copying all the tabbed pages, highlighting the text referencing part replacement, time-based maintenance, and operating activity and then organizing and binding the material completes the PM documentation. Vendor manual control reference information and index are also helpful to establish PM input traceability (*see* Fig. 3–3).

Any reference material can be treated the same as a vendor manual, depending on the credibility of the source. Many useful sources are available, including websites (see Noria's lubrication site), regulatory sites (see the US NRC's site—*www.usnrc.gov*), or historical engineering analysis. Reading white papers, technical papers offered at conferences like the ASME's Power Generation Conference, the Society of Maintenance and Reliability Professionals' Annual Reliability Conference, and the American Nuclear Society's Utility Working Conference are excellent ways to stay abreast of critical industry emerging-failure problems. The selection of any reference material supporting a failure mechanism's PM task depends on the reviewer. Some sources are outstanding; others offer methods and techniques that should be treated with caution until validated by experience. The review of test and assessment techniques like partial

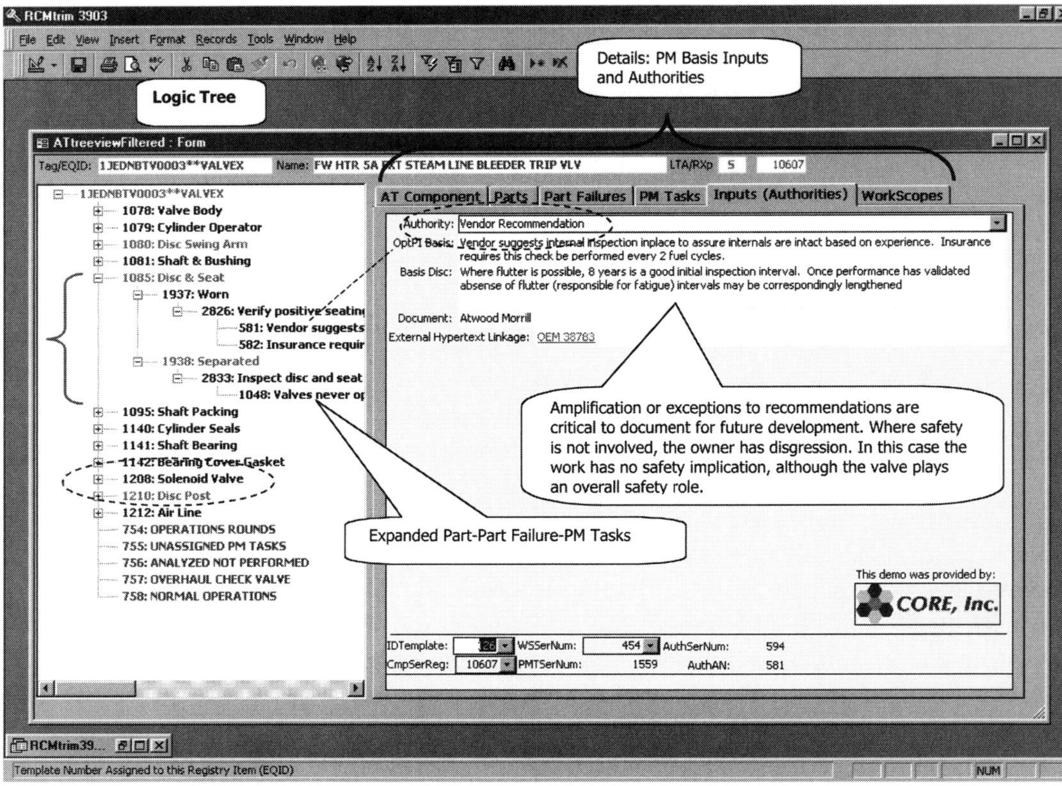

Fig. 3–3 Generic Template PM Tasks and Basis

discharge (PD) monitoring or ultrasonic analysis is very specific to symptoms. Basis information should be supported with actual analytical experience. A great deal of book theory doesn't hold water in practical applications often because lab conditions cannot be reproduced in the field.

Interviews with experienced operators and craft are another source of basis documentation that should get recognized when an analyst reviews and validates their methods and documents them for future use. Frequently operators and mechanics are familiar with symptoms by virtue of their floor presence—which engineering staff does not have. Capturing these nuggets assures that techniques aren't lost should an early retirement program, for example, remove 25% of the maintenance force at a facility.

Building binders of vendor technical information by equipment types—vertical pumps, large motors, check valves, and so forth—can provide an easy way to consolidate data that is very focused on specific classes of equipment and their maintenance methods. Binders can be organized around vendors, classes, or other meaningful criteria. These binders provide a repository of information for reliability engineers and maintenance technicians and are the prime source of information for their work. They provide the text document bridge supporting the electronic technical templates.

In the end, the reliability expert wants

- reference material on the equipment: component class, type, vendor, source documents
- representative picture or schematic of the equipment to gauge size, types, installation etc.
- index of O&M/vendor technical manual resources used
- design sections describing the way the equipment works
- operations sections describing how to operate the equipment
- maintenance sections describing how to troubleshoot and maintain the equipment
- technical notes for modifications or problems encountered that have been addressed

With these materials, the reliability engineer has the tools to perform the next step—identification of likely dominant failure modes.

Dominant failure modes represent the way the component fails to perform its design function. Pumps deliver flow at pressure (head). Failure to pump water describes a component failure mode. Generic component level failure modes describe neither parts nor specific requirements, which are specific to the application. Parts failures cause component functional loss, and this functional loss is what a craft person must eventually diagnose if a pump doesn't pump. Components divide into parts, the next level for analysis. Expressing components as parts prepares for the next step—identifying the dominant failure modes by their associated part failures that affect component functional performance. For clarity, *failure mechanisms* define part dominant failure modes identified by engineering causes. Failure mechanisms are the combinations of parts, with the engineering causes like erosion wear that result in component failure. Maintenance finds and corrects part failure mechanisms to restore functionality. (*see* Fig. 3–4)

Expressed dominant failure modes depend strongly on service application and environment. How often does the component run, start, and stop? How do operators treat the component? Is equipment inside or out, temperature-controlled or not, humidity/dust controlled or hot, dry and dirty? These factors all influence equipment service life.

Risks factor into PM tasks selection. Most equipment comes with instrument packages that assure high-risk hidden failure modes are evident and can be managed. Alarms are provided for high-risk failures, for instance; they may require calibration or routine operational performance test.

Engineering skills support identification and quantification of equipment parts failures as failure mechanisms and relate those to the failure modes seen by operators. Engineers often seek complete failure mechanism listings, but most equipment has only

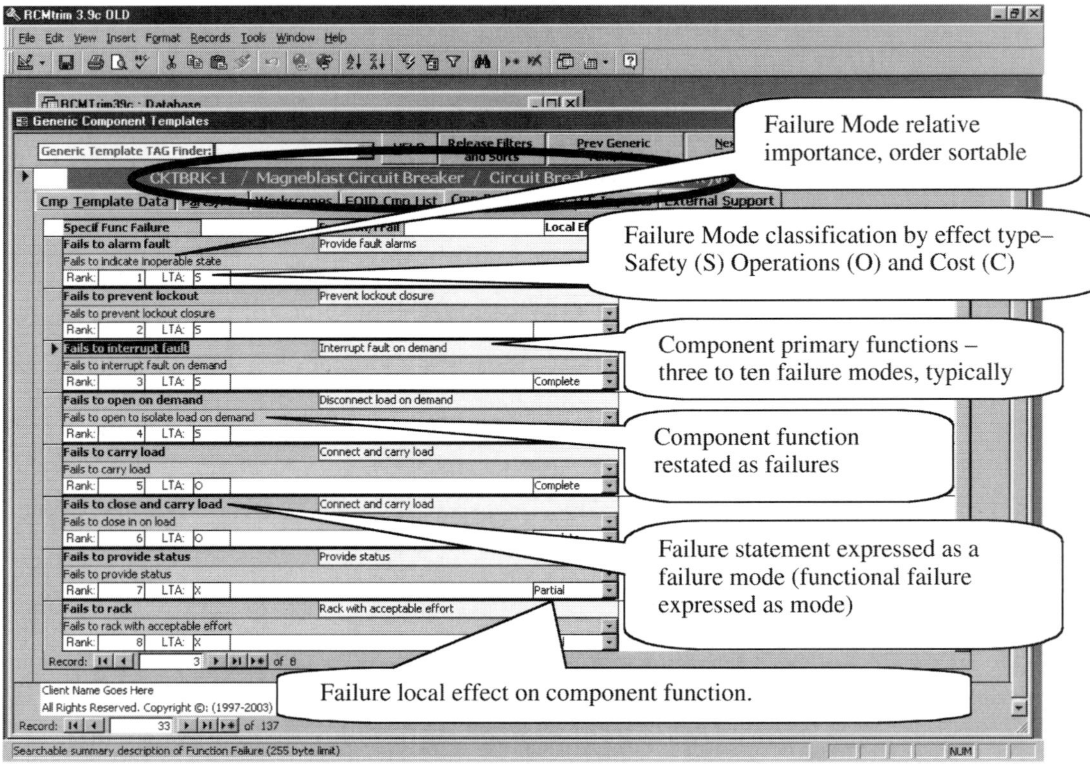

Fig. 3–4 Component Functions (4kV Breaker)

one or two dominant ones. For example, motors predominantly fail by bearing loss, wiping windings from interference, or from motor insulation failure due to aging. Industry statistics indicate that 45% of motors are damaged from external bearing-caused wiping and 30% from insulation aging. That is a startling statistic indicating that most motors never reach their potential winding-based life due to bearing failure! Improving motor lifetimes means monitoring bearings and identifying faults in advance of winding damage. The intervals between detecting bearing damage and replacement must be short enough to head off the failure but long enough to allow bearing-related failures to develop. Determining this interval requires combined knowledge of bearing aging deterioration and the value added from avoiding the motor damage. For large motors this is a desirable goal—to avoid cost and performance loss.

Ultimately, experience determines those failures that occur frequently. Where failures have never occurred but are suspected, part opportunity samples can be collected and examined to confirm or reject the occurrence of the failure mechanism. Clearly, vendors can't do this work unless engaged by the owner to provide aftermarket maintenance and support services. Where companies have large populations of the equipment in service (or modest populations of equipment with years of service), it's not difficult to collect failure statistics. This actuarial data determines lifetimes, failure types, and potential technologies, and the tasks to manage failure. The collection and reduction of actuarial data is hard work. It requires statistical knowledge and a willingness to extract work order failure information and reduce it to Pareto block diagrams to display the relative frequency of failures for a given class of equipment. Because of the rote time and intense effort involved performing these tasks, some have sought to simplify and streamline the failure-reduction process. A few have gone so far as to discount entirely the value of failure-data collection.

But real data keeps engineers' feet on *terra firma*. Failure to relate PM task recommendations back to real data risks development inefficiencies. No failure events are retold as frequently as the most recent one! Insisting that all PM task inputs are based upon data controls personal biases, memory lapses, and many other subjective factors.

Interviews provide an acceptable substitute for data and should be considered when data is not available. Interviews are easy, fast, and provide a wealth of operator and maintenance perceptions when there's no other sources available—data or otherwise. Prepared interviews are more productive than informal ones. Offering ideas about problems or making provocative suggestions about how to perform maintenance opens craft worker discussions. Once discussions begin, the information freely flows! Talking about failures is a love of craft workers, for correcting failures is their work. Interviews can explore suspected trends and examine subtle problems that data suggest. The inadequacies of parts, the overall value of original equipment manufacturer hardware parts compared to competitively available alternatives, and other equipment and maintenance issues readily surface in such small group discussions. (*see* Fig. 3–5)

Once the dominant failure modes are developed, a nominal service interval must be found. Fortunately, at the generic template level, it is not necessary to specify any exact replacement intervals or even task limits. Aging parts replacement or their equivalent on-condition discovery tasks that identify deterioration onset can be picked at the time of template application. Providing a complete set of well-developed and thought-out PM alternatives is what's needed.

For example, at one plant the discharge head of a small electrically driven startup boiler feed pump could be trended under similar startup conditions. A steady decline in the pump discharge head pressure at flow into the boiler drum could be tracked and trended over time and compared to the minimum pressure required to charge the drum adequately to steam the boiler. This situation suggested a classic on-condition task: measure the parameter, trend the measurement against the target performance limit, and plan to perform maintenance when the target exceeded the limit based on projection from the last measurement. The analyst needs to know the pump performance test alternative is available for the case where continuous wear out is not applicable.

Setting service intervals requires insight into the deterioration processes, intuition, and willingness to take risks. Unless a failure has safety implications, any limit is better than none; perfect selection is not essential. Where safety is involved, vendors often specify limits, or station licenses specify requirements based on safety consideration. When AEs specify equipment for safety applications they put their requirements in system design descriptions.

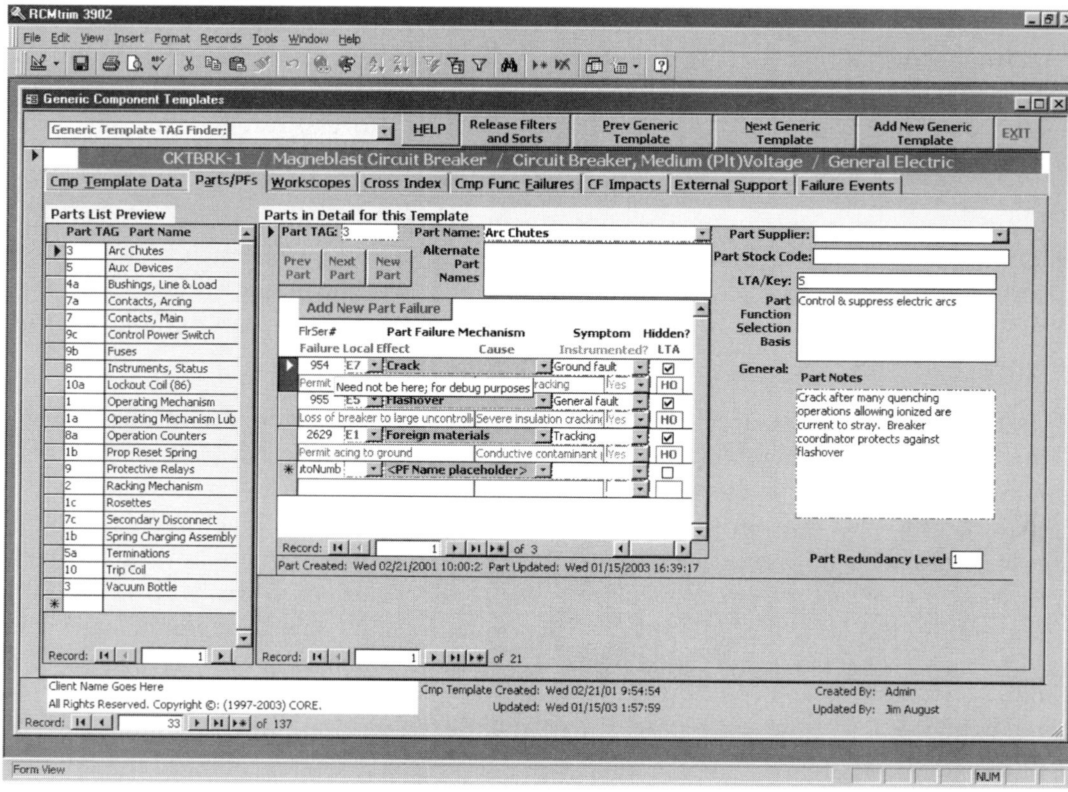

Fig. 3–5 Generic Template Part Failure Hierarchy Showing Dominant Failure Modes

Common problems

When building generic templates, too much engineering familiarity causes as many problems as too little. Familiarity also leads to allocating too little time to adequately research equipment failure history or references. Often with familiarity comes the assumption that equipment and problems are familiar and known. Failure analysis on even the most common equipment provides concrete failure distributions that always tell a story.

Component risk is difficult to establish outside a physical context, and generic templates lack context. For example, a breaker may inadvertently trip and drop a bus, but unless the lost bus powers control rod drives at a nuclear plant, the consequences are likely limited to load loss. Design addresses the losses that potentially cause plant safety issues. Control rod drive power can be redundantly supplied. Automatic throw-over bus tie switches are likely to be installed to switch to alternate power on loss of load within a half-cycle where it is critical to do so. All this detail is contextual, beyond the generic template.

From a generic template, one can infer only generic information. Equipment risk context can't be established until the template used to model a plant-installed component is applied. At that time risk can and must be accommodated. Some component-specific parts specifically manage component-level risk.

For example, large rotating fans have built-in vibration trips for excessive vibration. When a preset limit is exceeded, the fan trips off to protect against possible missile ejection from the large rotating mass. Rotating inertia is so great that an imbalance can overload the bearings, ripping them apart. Failure to acknowledge the vibration trip's safety role introduces a hidden failure that could injure or kill. Vendors therefore provide the trip as part of the skid package and print strong language concerning the importance of the vibration trip in their O&M literature.

Manufacturers know previously identified, catastrophic failure modes for their equipment. Vendor manuals provide their readers with critical, rare, one-time failures that contribute to real equipment operating risk; they amplify cautions with bold warnings about risky tasks and warning alarm features and instruments. Supporting manuals, industry-operating experience, and industry forums are great sources of operating risk knowledge, particularly for catastrophic events. Word of mouth information exchange always occurs where losses are high and casualties may have occurred.

High-loss events often involve secondary failures. By studying high-loss events reported by trade organizations and insurers, one gradually appreciates the need to understand the integrated design basis for complex equipment. Mature-design equipment's catastrophic events often involve a combination of missed warnings and operating ignorance. The challenge is to make sure alarms and operating limits are maintained and to ensure that operators understand the operating risks fundamentally associated with equipment and design features intended to keep operations within the design range and that they follow safe operating procedures.

Annual insurance loss reviews are another useful learning-development tool. Event descriptions illustrate common problems including ineffective protective devices, warning systems out-of-service for safety or fire, and other high-risk industrial equipment events. These multiple-event, secondary failure occurrences provide awareness of protective devices and safety features for fuel, fire, and electrical systems that commonly support many industrial enterprises. Special trade groups and organizations like the Institute of Nuclear Power Operations provide similar industry-event dissemination services.

Alternatives

Developing templates from scratch is intense. It needs to be performed several times to learn fundamental steps and to gain basic proficiency. For some equipment it may be easier to borrow previous template parts and edit them.

For example, a main turbine template could be edited to serve a small boiler or reactor feed pump turbine or even skid packages (a Terry turbine, for example) with modest work. *Cut and paste* template development becomes even more attractive when encountering variations on a basic model or a new manufacturer's design that is physically similar to one already modeled. In this case, it's a mere matter of copying the parts list, functions and failure modes, and failure mechanisms, and doing some light editing to create a fundamentally new template. Being able to do these templates quickly is necessary in production RCM projects.

Engineers using these techniques can build site-specific templates using pre-existing, customized maintenance processes and methods to address definable failure modes. For example, one site may have thermography cameras available to perform equipment thermal surveys; another may have digital-reading temperature guns. The activities might be substantially the same, but checks and interpretations would have to be tailored for each. Rapid template modeling can create appropriate, on-condition, predictive tasks suitable to any site.

Functions and Failure Description

Component failure modes

Operators interpret equipment condition knowing common equipment failure modes. Equipment (component) failure modes are one level up from the engineering failure mode interpretations used by mechanics and engineers. Engineering failure modes are referred to as part failure mechanisms or just failure mechanisms.

Experts may be needed to discern failure details once symptoms are discovered. Leaks, noises, smells, and vibrations—symptoms of integrated part failure—could result from a variety of specific problems. Imbalance, misalignment, binding, and other part deterioration failure mechanisms cause vibration. Failure diagnosis looks inside the box; the determination that a failure mode is apparently in progress, based on symptom, leads to diagnosis. Leaks can be through-wall, stem packing blow-by, gasket cracking, or from many other causes. An operator need only *discover* failure. Experts help diagnose failure mechanisms for correction as condition-directed maintenance.

Component failure modes are function failures at the component level. Their development follows the same outline as developing system functions and function failures. Expected component output(s) are listed first. For example, valves are expected to contain fluid, operate freely on demand, and isolate by the seat. As with systems, passive requirements like *flow freely open* are easily overlooked. The main difference in developing component function statements for generic templates in the specific function versus for specific templates is that numerical requirements are commonly omitted. This convention allows interpretation of those requirements to be integrated with systems. Until templates are specifically related to a system, performance requirements lack context for diagnostic application by the maintenance worker (*see* Fig. 3–6).

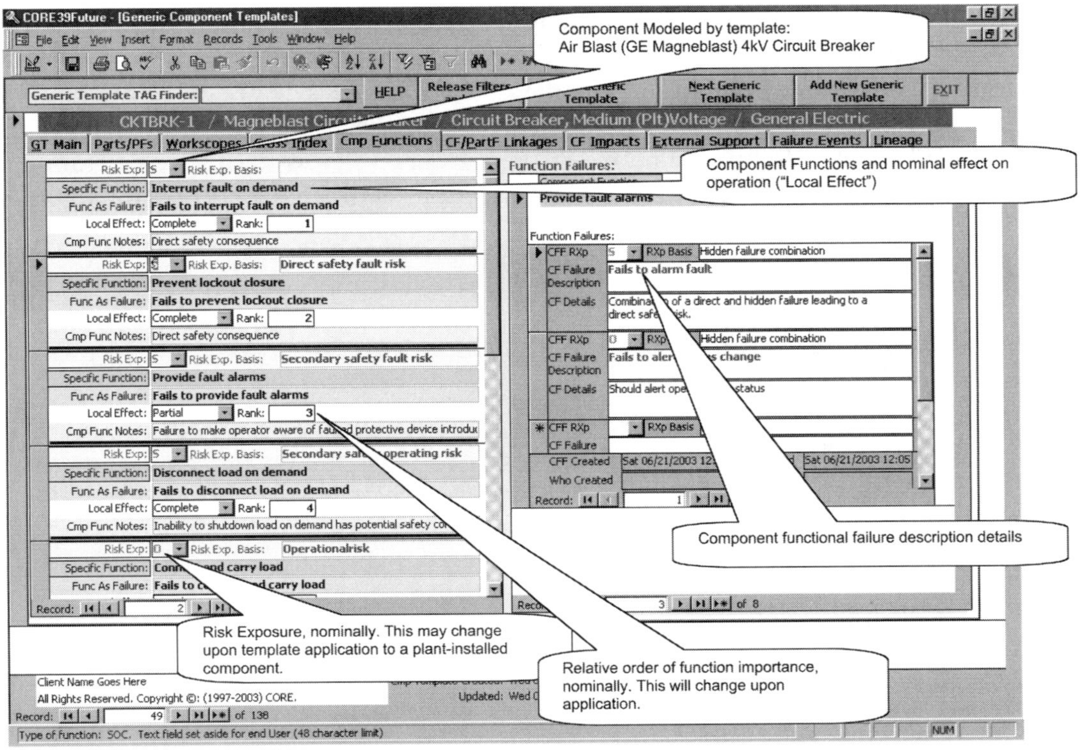

Fig. 3–6 Component Failure Modes

Part failure causes

Engineering failure modes are part failure causes. Metal fatigue causes fatigue cracking. Material erosion causes loss of strength and yielding. Parts fail from the combined effects of stress and time. The engineering failure cause identifies the physical mechanism at work. This suggests an aging time parameter of interest, as well as symptoms that identify the failure process at work. For fatigue, an aging life of 1.1E06 cycles at 20% of rated yield load defines a life parameter. Fatigue process cracks are end-of-life symptoms.

Engineering failure-cause identification is important to establish estimated life and to define the failure symptoms that will successfully identify on-condition maintenance programs. Fundamental engineering failure causes are quickly learned by degreed engineers, based on college course presentation. Non-degreed technicians must learn new failure descriptions and mechanisms. Becoming expert in any specific engineering failure process (like stress corrosion cracking) easily takes a lifetime. Fortunately, experts are available as consultants. For scheduled maintenance program development purposes, it is advisable to seek failure recognition and identification at the 90% competency level. It is also recommended to specify failures with symptoms that are easily recognizable for fieldwork.

Providing acknowledged failure symptoms develops failure mechanisms for field WO workscopes. Failures that are not within expected lifetime ranges need further engineering evaluation. Premature cracking could indicate a new failure mechanism such as stress cracks and not the anticipated one—fatigue. Engineers and technicians can readily recognize common failures at high accuracy levels with specific training. Reasonable accuracy calling failures, not perfection, is all that's needed for maintenance program success (*see* Fig. 3–7).

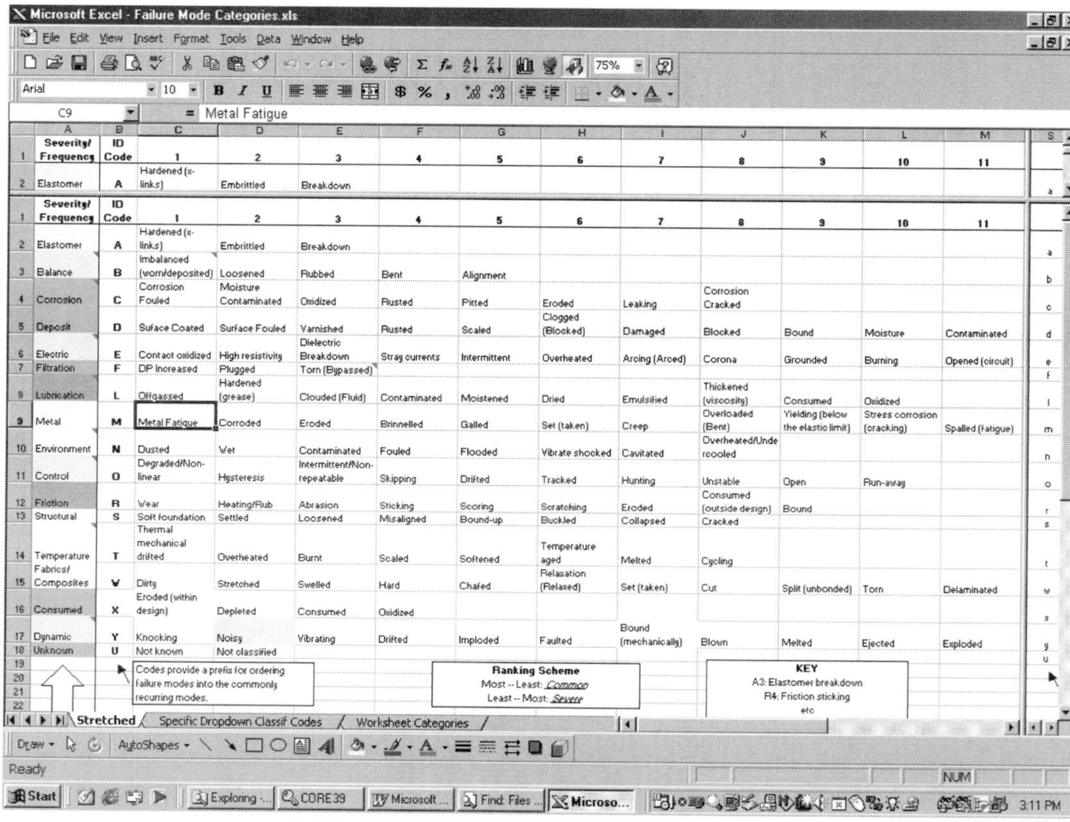

Severity/ Frequency	ID Code	1	2	3	4	5	6	7	8	9	10	11	S
Elastomer	A	Hardened (x-links)	Embrittled	Breakdown									a
Balance	B	Imbalanced (worn/deposited)	Loosened	Rubbed	Bent	Alignment							b
Corrosion	C	Corrosion Fouled	Moisture Contaminated	Oxidized	Rusted	Pitted	Eroded	Leaking	Corrosion Cracked				c
Deposit	D	Suface Coated	Surface Fouled	Varnished	Rusted	Scaled	Clogged (Blocked)	Damaged	Blocked	Bound	Moisture	Contaminated	d
Electric	E	Contact oxidized	High resistivity	Dielectric Breakdown	Stray currents	Intermittent	Overheated	Arcing (Arced)	Corona	Grounded	Burning	Opened (circuit)	e
Filtration	F	DP Increased	Plugged	Torn (Bypassed)									f
Lubrication	L	Offgassed	Hardened (grease)	Clouded (Fluid)	Contaminated	Moistened	Dried	Emulsified	Thickened (viscosity)	Consumed	Oxidized		l
Metal	M	Metal Fatigue	Corroded	Eroded	Brinnelled	Galled	Set (taken)	Creep	Overloaded (Bent)	Yielding (below the elastic limit)	Stress corrosion (cracking)	Spalled (fatigue)	m
Environment	N	Dusted	Wet	Contaminated	Fouled	Flooded	Vibrate shocked	Cavitated	Overheated/Underooled				n
Control	O	Degraded/Non-linear	Hysteresis	Intermittent/Non-repeatable	Skipping	Drifted	Tracked	Hunting	Unstable	Open	Run-away		o
Friction	R	Wear	Heating/Rub	Abrasion	Sticking	Scoring	Scratching	Eroded	Consumed (outside design)	Bound			r
Structural	S	Soft foundation	Settled	Loosened	Misaligned	Bound-up	Buckled	Collapsed	Cracked				s
Temperature	T	Thermal mechanical drifted	Overheated	Burnt	Scaled	Softened	Temperature aged	Melted	Cycling				t
Fabrics/ Composites	W	Dirty	Stretched	Swelled	Hard	Chafed	Relaxation (Relaxed)	Set (taken)	Cut	Split (unbonded)	Torn	Delaminated	w
Consumed	X	Eroded (within design)	Depleted	Consumed	Oxidized								x
Dynamic	Y	Knocking	Noisy	Vibrating	Drifted	Imploded	Faulted	Bound (mechanically)	Blown	Melted	Ejected	Exploded	y
Unknown	U	Not known	Not classified										u

Fig. 3–7 Engineering Failure Causes and Mechanisms

Critical failures

Some failures pose very high risks, which may affect safety. Large rotating masses in turbines and compressors or fans pose missile ejection hazards under overspeed or imbalance conditions. Pressure vessels pose catastrophic rupture risks. Other chemical processing, transportation, and refining equipment incur similar risks. Sometimes risks arise from inoperable safety subcomponents on equipment. Equipment intended to mitigate risks can potentially become inoperable.

Steam turbines have overspeed potential, so overspeed trip protection removes the credible risk of steam supplied in the absence of load. Large 15-foot induction draft fans credibly develop imbalances that shed parts. These have vibration trip limits to protect against credible fan wheel missile ejection. Smaller turbines have cowlings to enclose and capture missiles. Similarly protected equipment has protective trip devices incorporated into its design. Overspeed and vibration trips are two such devices to protect against otherwise catastrophic, credible events. Should trip devices fail—and they are in continuous standby service during equipment operation, the failure they are intended to protect against is neither protected nor apparent; and catastrophic consequences can occur. The design-protected failure becomes unprotected.

General turbine trip logic relay failure (simplified) could trip the generator without a corresponding turbine trip. ID fan wheel crack propagation could cause imbalance without a corresponding fan imbalance wheel trip. Overpressure of a boiler due to excess firing without a safety valve lifting could cause catastrophic mechanical pressure wall failure. Each event illustrates MSG-3 hidden failure with an associated safety logic criterion.

Does the combination of a hidden function failure and one additional failure of a system related or back-up function have an adverse effect on operating safety? When the answer is yes, the case typically stays with one of the following patterns:

- The hidden function is provided by the standby protective device (whose failure is the hidden function failure).
- The additional failure initiates the normally-protected catastrophic event, whether it is over-admitting steam to the turbine, not tripping an unloaded turbine, or failing to shutdown a severely imbalanced fan.

Practically, suppliers of equipment with known hazardous operating regimes or failures provide or specify safety devices. Codes, OSHA standards, statutory laws, or workplace hazard rules typically address these devices as well. Critical safety devices are not difficult to identify with experience, although rule interpretations vary and can be the source of endless discussion and even litigation. Rule interpretations aren't needed when things work, rather they're required when devices don't perform as expected.

Interpretations of grace periods that allow for orderly shutdown, compensatory measures, or other special requirements are bones of contention that introduce risk when devices misbehave. It's hard to shutdown major production lines or equipment for a failed safety device. Interpretation of failing devices can be complex, as in the case of a safety valve that fails to lift within limits. It is a lot easier and less stressful to know equipment works and have exact guidance and specific limits before failure occurs. Interpreting equipment failures in advance as RCM failure *local effects* provides previews of necessary plant actions. This is when likely actions and consequences should be thought through. This improves awareness of failed device operating consequences and increases commitment and assurance that they are correctly and adequately maintained. This is a significant contribution of RCM, though not one consistently acknowledged.

During critical equipment parts failures analysis, failures with direct potential to kill or injure, in particular, should be sought for associated risk mitigating device(s). At this time, it's easy to see hidden function and direct safety function failure pairs in the design. (An MSG-3 pair is a critical protected failure with a protective device subject to hidden function failure.) Where found, corresponding hidden function failure modes get elevated in importance and must be addressed by scheduled maintenance to reduce hidden failure probability to an acceptable limit. Usually the task is a simple failure-finding test. Identifying hidden failures is the single most valuable RCM risk control practice for most industrial facilities (*see* Fig. 3–8).

For skid-mounted or wholly supplied equipment, hidden functions/failures are evident from vendor literature. This integrates equipment risk and maintenance requirements. For AE-supplied engineered equipment, failure strategies are incorporated into design but may be subtle. Anticipated failure consequences and design-mitigating features are buried in *engineering system description* text discussion, rather than printed O&M manuals.

In industry, practical operating risk occurs when protective devices fail and fall out of operating perception. Forgotten and fallen by the wayside out-of-service, any appreciation for the failure control features and the added value represented can be totally lost. Loss of awareness of failure control features and their value controlling practical operating risk occurs in plants over time. In these cases, the benefit to operating reliability and the cost benefit realized by rediscovery and correction results in substantial performance gain at minimal cost. In any event, analysts should seek hidden function failures from their associated protected failure events, quantifying the protected failure avoidance benefit. This good RCM practice assures the plant gains all benefits from the functional review.

Where no corresponding protected failure or control function can be identified, an instrument may have solely status function. Finding and associating critical part failure/critical instrument pairs requires learning to identify standby and hidden safety function equipment. Learning instrument value contribution comes from risk analysis (*see* Fig. 3–9).

Instrumentation and controls

Instrumentation typically plays one of three roles: control, alert/trip, or status. Control is usually evident although fully redundant, dual-control highways in modern DCS systems offer considerable built-in redundancy. Modern, state-of-the-art control systems provide excellent self-diagnostic capabilities. Operational system failure-to-control is usually evident in single-loop control loss. Of more reliability and cost interest are the other two roles: alert/trip functions and status. Main plant controls are

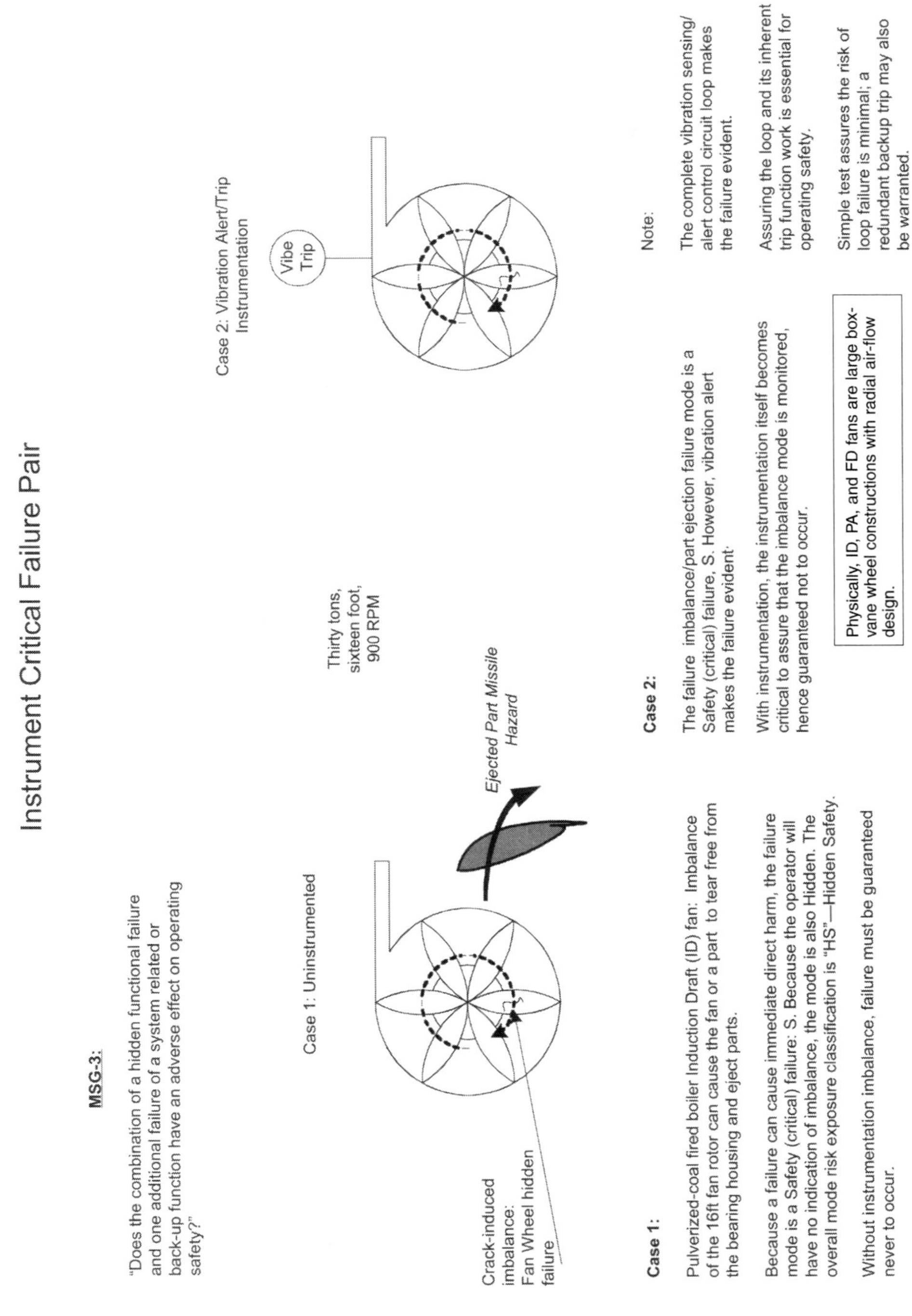

Fig. 3–8 Instrument Critical Failure Pair

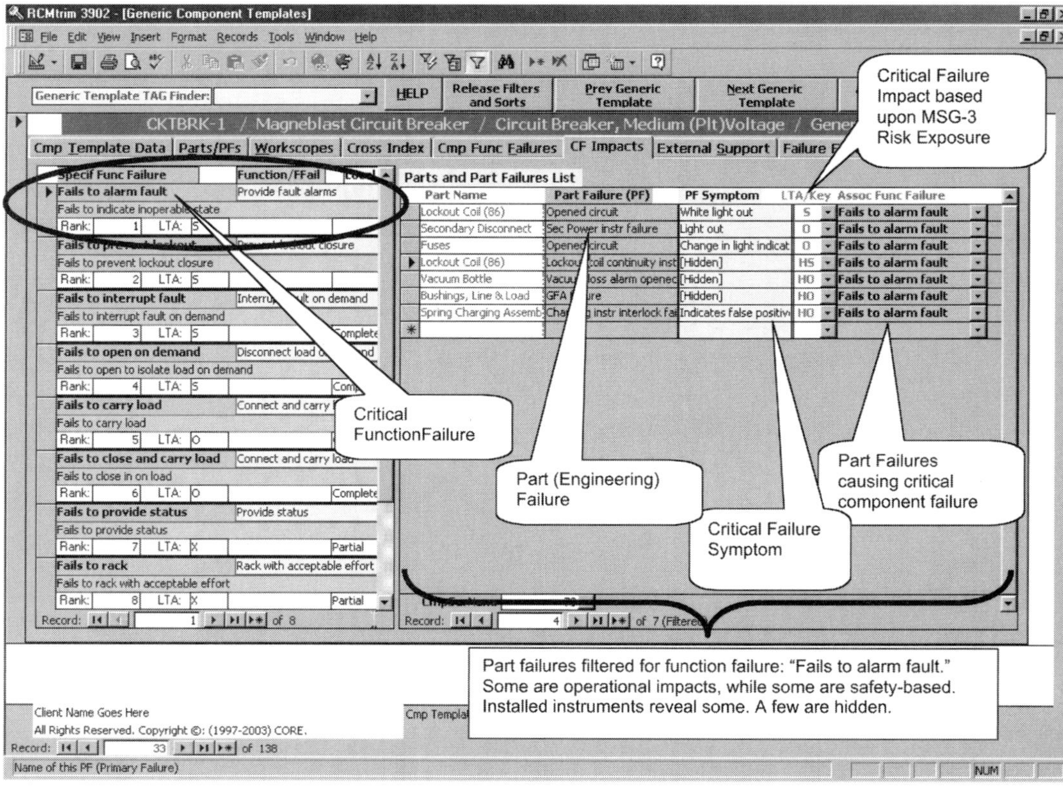

Fig. 3–9 Critical Component Failures

a separate, unique system and can be analyzed as systems; most equipment skids provide local instrumentation and control, which may act alone or integrate with plant control system interfaces.

Instruments in alert/trip loops operate in continuous standby, unlike active controls, which fail to show control loss. Stand-by control failures will not be evident until a demand event occurs when functionality loss fails its intended function. Loss of hidden failure functions only becomes evident at the moment of need, and that's too late. These devices often combine redundancy in application. For high-risk failures, a hidden-failure risk mitigation strategy performs a periodic operational test. For a period following a successful test, *in-service* probability remains high. The operable state decays with a characteristic exponential curve behavior. At some later point, another test is appropriate to validate continued functionality. Frequency of testing determines the probable risk of discovering an existing failed state. By adjusting task time interval, this risk can be managed.

Henry's Proposition

On-condition hidden failure monitoring confirms that hidden function remains active. Testing for the hidden function forces a hidden failure to reveal itself. Thus, performing scheduled maintenance tasks forces a hidden function to become evident.

(This is called *Henry's Proposition* after its proponent.) Continuously monitoring functionality of otherwise hidden functions also makes the function failure evident, as is the case in electronics circuitry and control. With continuous functionality monitoring, hidden failures are made evident by alarms. Routine scheduled maintenance performance has a similar hidden function-revealing effect. Revealing lost hidden functions is the generalized role of a scheduled maintenance program.

The continuous monitoring function itself is probably hidden and should be validated periodically. In this manner there is a chain of hidden functions, each revealing another's failure. Overall risk depends on the reliability of each link to reveal failure and in time required to restore functionality. These multiple layers of defensive depth could be programmed into a PC or microprocessor, but each succeeding level gets ever more complex to analyze from a reliability perspective. At some point, simulations become more attractive!

Parts Partition

Risk exposure

Within a template-modeled component, different part failures have different consequences. Some parts immediately fail the component; others do so in a progressive manner.

Directly failing a component requires single failure. Components commonly have multiple functions, so part failures can have different consequences. Some part failures cause explicit conditions and warnings; others require interpretation or secondary tests. Selecting a failure mechanism limit provides an unambiguous condition to determine the failure presence. Upon performing the condition check, exceeding the limit initiates condition-directed maintenance. Explicitly reviewing and setting limits is difficult engineering work. Many facilities have never taken these steps systematically at an engineering level.

Historically, failure presence determination depended on skill of the craft. Setting objective limits makes PM tasks individually actionable. This important step reduces failure management to specifications. Setting limits is imperfect, but once set, known condition-directed maintenance can be unambiguously performed. With repeated efforts to establish limits, competence setting limits grows, and initial limits set can be reset or redefined based upon that experience. Continuous improvement demands continual development, setting, and resetting of limits.

Risk partition

Part failures cause component failures that lead to system function failures. Fundamentally, system failure begins with the part failure mechanism. To manage system failures, ultimately the organization must understand and manage risk at the part-failure level. Preventing part failures requires understanding fundamental engineering failure mechanisms. This allows detecting part deterioration in time to head off failure before it compromises system functions. This establishes the PF-F interval, which establishes the lead-time for an on-condition/condition based maintenance task pair.

All failures are not created equal; some pose more risk. Mature designs effectively remove risk from the majority of their failures, specifying parts replacement or reworking of tasks. Identifying part failure with risk assists the development of analysis of component failure, and ultimately of system failures. Providing the nominal part failures for standard out-of-the-box components allows the development of tests, likely system failure impacts, and most probable high-risk dominant failures when the generic template is applied to a real context. Relating system failure risk down to the level of part failure mechanism retraces the failure event chain to its point of origin. That action helps quantify and eliminate risk (*see* Fig. 3–10).

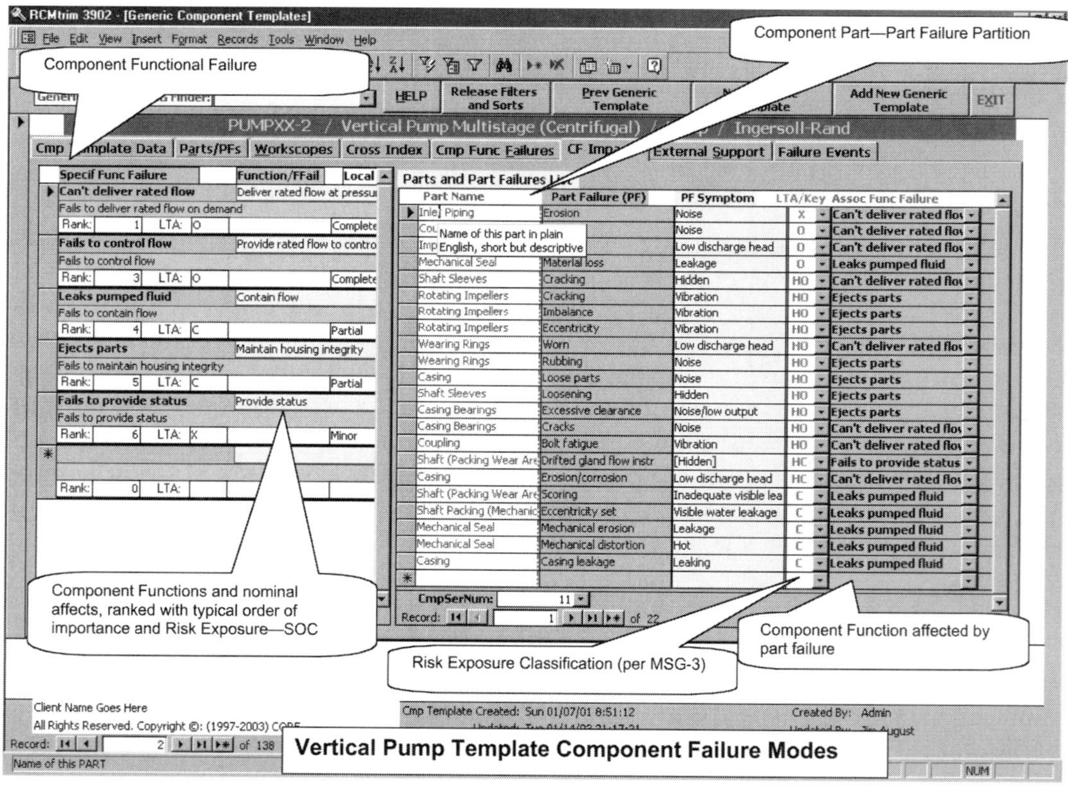

Fig. 3–10 Part Failures with Component Failure Modes

Instrumentation and controls

Much procured equipment is skid-mounted, so many components have integral monitoring and control packages. The nature of these packages—their isolation or integration with plant control schemes and component support roles—is part of the component analysis. Monitoring and alert are two basic functions that are always provided. *Alert* is provided as a warning while trips detect and interdict risky failures. These include safety and operational functions, as well as system protection functions.

For a skid-mounted instrument package, determining the role of each warning or alarm indicates the relative merit the item warrants for scheduled maintenance. These components provide equipment protection and, therefore, have a cost basis. The exceptions are clear from vendor literature or AE installation. Devices with safety functions must be treated with the same consideration as safety-equipment at the plant level. Where safety features may be easily overlooked or are unclear, vendor O&M literature indicates the vendor's test recommendations and suggests the safety roles, often directly. AE system integration engineers can use any equipment to perform a safety role. Their descriptions typically acknowledge those cases that are unusual. Very often, control and alarm packages include safety functions.

Most skid-mounted equipment provides equipment status. Unless clearly indicated otherwise by cost or production support, this instrumentation doesn't benefit from scheduled maintenance. Practically, a few useful indicators or status controls will benefit from routine calibration. (For these indicators, scheduled calibration is better than operators constantly requesting calibration!) For many plants, however, removing status instrument calibration provides a significant reduction in scheduled maintenance. Restricting calibrations in the unit outage frees up significant resources for other tasks.

Calibrations in copy composite (clone)

Cloning an existing template provides a complete model for fast editing. This is a quick way to develop a similar, new template. Cloning an identical complex template for similar equipment (like another model from the same basic design) or for another manufacturer's equivalent design provides an analysis force multiplier to model similar equipment generic templates. The need for a copy-and-paste routine becomes very apparent when building many similar templates. With spreadsheets, analysts quickly find they are copying full sheets to speed new template development. As a practical method to develop a fundamentally new design, but reusing a well-developed model as the foundation, template copying easily extends to database application designs (*see* Fig. 3-10a).

Resources

Resource requirements for PM tasks should identify the approximate time required for a proficient worker to perform the task. Times should presume tools and materials are available, although these should be separately identified as PM task requirements. Tools and special skills, like craft skill hours, are resources, so they are generally available performing work. Special required parts should be evident, based upon type of PM task and potential requirements. For example, on-condition testing for tube wall thickness may require having tube plugs available to plug thin tubes. Hard-time replacements for filters are simply replacements, and every task performance requires filter replacement. The combination of task types (hard-time based replacement, condition monitoring, etc.) statistically determines parts usage and stocking requirements.

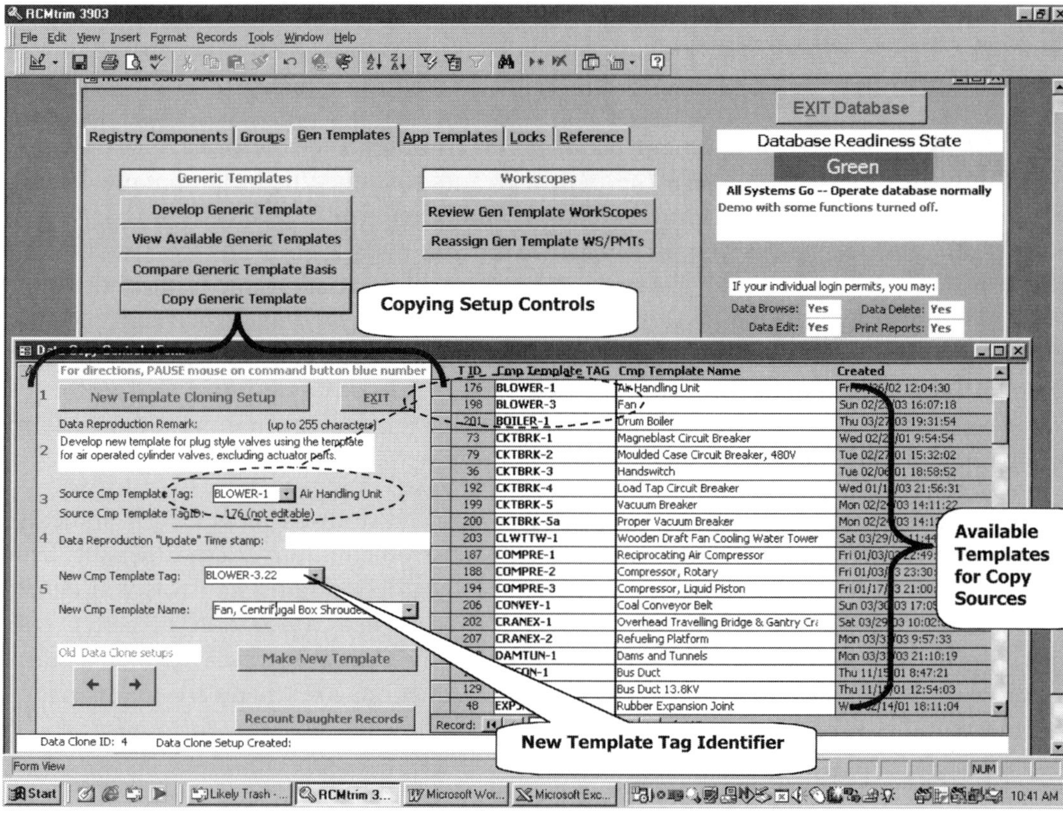

Fig. 3–10a Generic Template Cloning

Maintenance has never fully accepted cost accounting practices like manufacturing. This is partly because the tools have never been made available customized for *maintenance* like simple CMMS/EAMS subroutines. By deconstructing WO to the task level, the relative value of every contributing task can be weighed on merit and held or removed purely on that basis. Allocating cost to PM tasks reduces them to an objective, measurable component of work.

Basis

Basis defined

The basis for any PM task is the reason the task is worth doing. Indeed, basis answers the question "why?"

Traditionally before RCM, a task reflected vendors' recommendations, code, insurance stipulations, or, in the case of special plants, operating license requirements. Individual engineer or lead maintenance foremen made decisions to select one source of PM input over another. Usually, there's reasonable consistency of PM input among credible authorities. Occasionally PM inputs are in conflict, and a source-input hierarchy must establish the actual task selected to implement. Opinions and recommendations should be weighed cautiously; less input typically comes from craft workers actually working on equipment than secondary sources. Shops controlled PM programs and leaned heavily on craft to interpret and apply vendor and other recommendations. PM work orders were treated as loose guidelines.

Floor-based PM programs have been reviewed many times in many forums. They over-perform maintenance, doing too much work on non-dominant failure mechanisms for non-critical equipment. They perform work too frequently. They don't take advantage of standby service equipment's low aging-to-time work based upon usage. Performing intrusive work too frequently on hard-time schedules introduces *infant mortality*. Programs failed to manage equipment with risk concepts; the maintenance foremen or planners running the program lacked operational risk insights applied to a particular piece of equipment. These insights come from studying reliability, serving as an operator, or performing in a high-risk department, like instrumentation/controls or operations.

RCM's PM task basis is failure implicitly prevented by a task, tempered by credible technology, supporting analysis recommendations, and experience that indicates a PM task cost-effectively eliminates a failure mode. By developing RCM tasks around failures, engineers explicitly justify tasks by providing an inherent, implicit basis for all work, embedded in failure analysis. The analysis behind the task is the formal justification. This is the *implicit* basis (*see* Fig. 3–11).

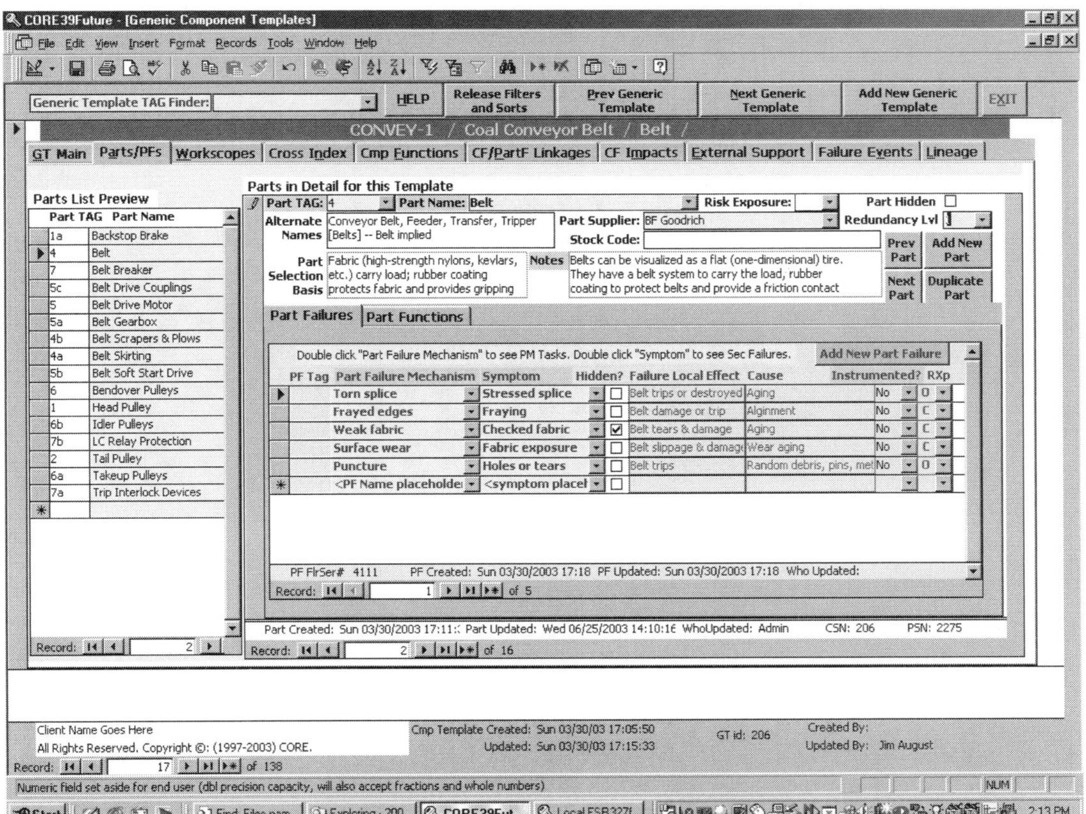

Fig. 3–11 Implicit Basis

The other basis is an *explicit* basis. In the past, codes, licenses, or even suppliers required PM activities. Laws like Title 10 regulating emissions controls and monitoring specified others. Station permits—wastewater reclamation, facility hazard management, and OSHA workplace rules for equipment such as overhead hoists, cranes and electrical equipment—all required tasks. These prescriptive requirements reached their limits with the very detailed operating license programs for nuclear stations. Force of law backed explicit PM task requirements.

In principle, explicit requirements related to PM tasks in work order workscopes tracked compliance directly and were retained for historical compliance verification. Demonstrating compliance is an explicit PM task basis, and the legal reason for the explicit basis. Where explicit requirements exist, maintaining an explicit basis assures program compliance.

Contrasting *explicit* basis with *implicit* basis:

- *explicit basis* relates PM tasks to rules, codes, and other legal requirements (failure prevented is implicit)
- *implicit basis* relates PM tasks to failure prevented (codes and rules, if any, are implicit) (*see* Fig. 3–12)

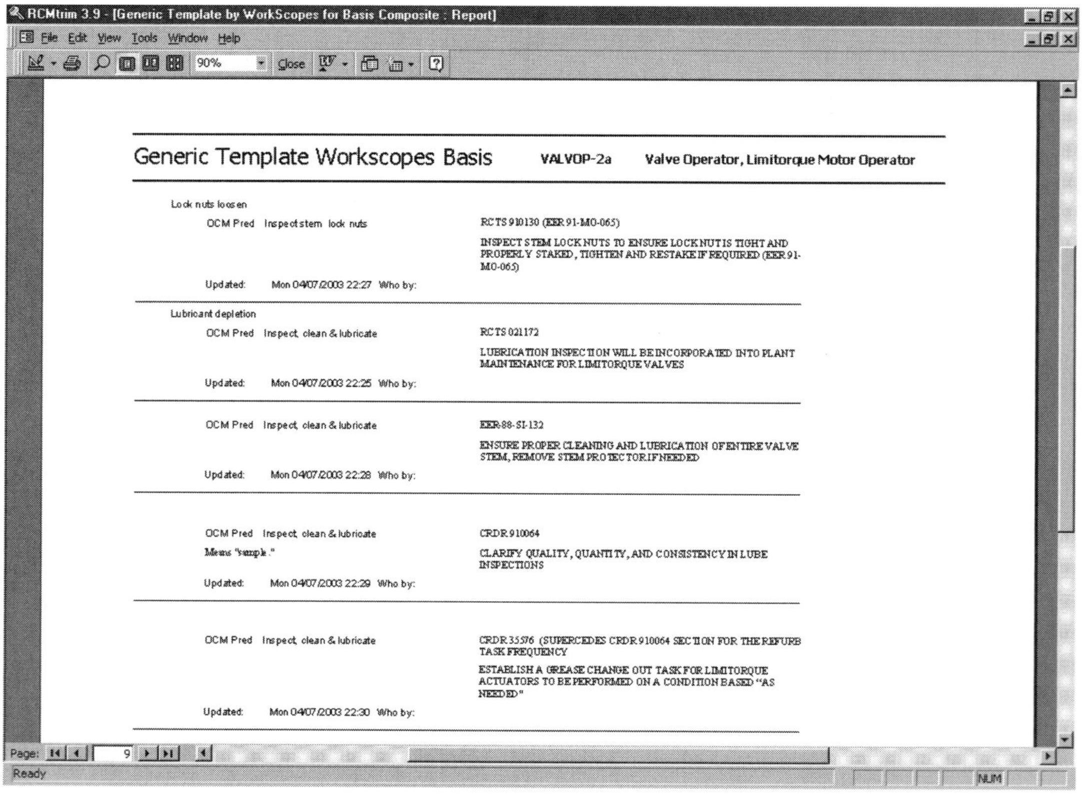

Fig. 3–12 Explicit Basis

Basis dilemma

A basis is simply a rational, supporting justification for task selection. Developing explicit bases is easier that developing engineering ones. The former is research, while the latter is analysis. It's always easier to do as you're told than to think. According to the old school, nothing is so important as the document from an authority that tells you what to do and authorizes you to do it.

Who, for example, would argue with the Wizard of Oz over the Lion's bravery (he had a Purple Heart!) or disavow the Scarecrow's brain (he had a degree!). According to the new school—the RCM School—documentation is interesting, but incomplete. Only the failure tie, an abstraction, creates a valid context for the PM. Ironically, this tie is what the explicit basis authorities sought to establish.

The preference any one person has for a basis reflects his philosophy and even ability to do basic failure analysis that creates a basis. Some industries are locked into aggressive, inefficient PM programs by combinations of basis interpretations and work commitments. Perhaps the most elegant use of MSG-3 and Nowlan and Heap's original RCM treatise is the rationale they provide that has stood the test of time in working through these deep issues in one of the most highly regulated industries of their time—the 1960s' airline industry. As so aptly pointed out in their 1978 publication, *Reliability Centered Maintenance*, the FAA federal rules that tell airlines to perform hard-time overhauls are still in the Code of Federal Regulations. They are simply not enforced administratively. An administrative policy white paper administratively superseded the application of that law, creating joint ownership responsibility for airline maintenance between regulator and licensee airlines.

Baseline program changes

Once a baselined PM program is in place, any program change should have measurable benefit to make the modification worthwhile. Developing a concurrent program change basis to justify program variations to the baselined program continues the program basis over time. All program changes are then documented and maintained for historical purposes. This change basis is helpful to maintain forward progress, avoid circular changes, pursue age exploration, and provide transition with personnel changes over time.

With more emphasis on PM cost management, a change basis can also document parts and process improvements that extend maintenance intervals, eliminate tasks, or convert expensive tasks (overhauls) to less expensive ones (condition assessment). These are the general characteristics of an RCM-based PM maintenance program. Time-based overhauls and refurbishment gradually shift towards assessment-driven condition-directed tasks.

Levels of basis

Just as basis answers why a PM task is performed, the same question could be extended to all levels of a template-based failure management plan. Why are the workscopes organized the way they are? Why are the intervals set at the limits they are? Why is this cracking failure mode considered but not that one? Why are some parts listed and others not? Why are certain PM technologies listed, while other perfectly adequate substitutes are not? (*see* Fig. 3–13)

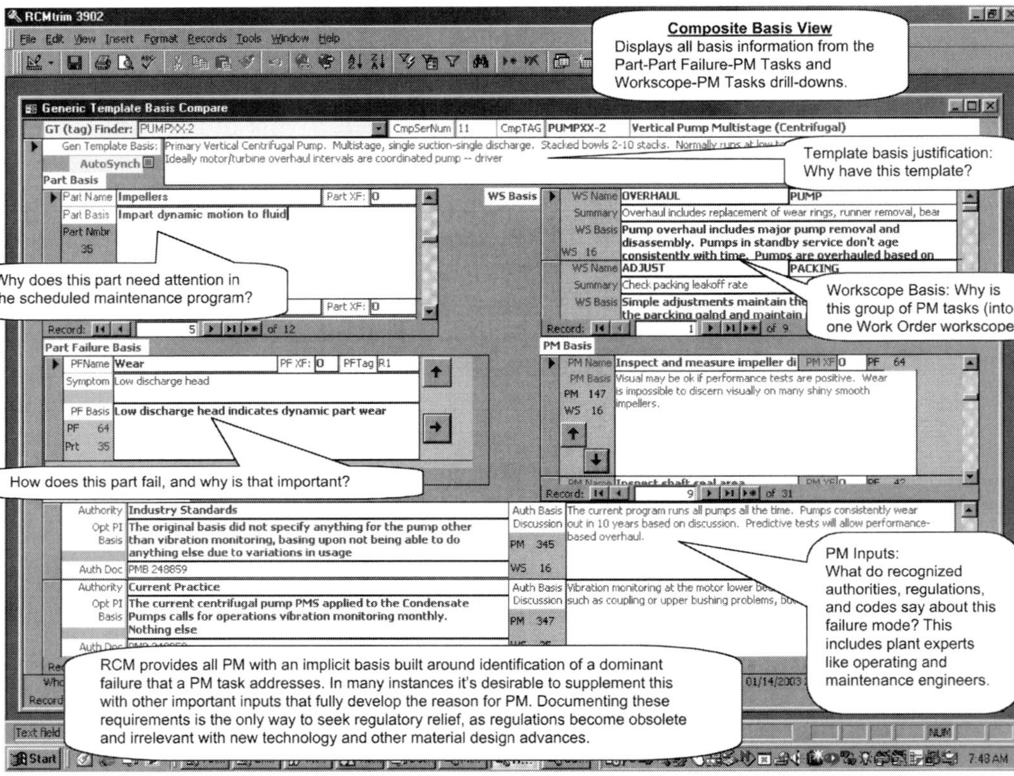

Fig. 3–13 Multiple Basis Layers

There is no limit to the number of times the question "why?" can be asked. A final authority such as an ASME Boiler and Pressure Vessel Code expert who could resolve the final questions of interpretation would be useful, particularly when conflicting requirements come into play. An expert could interpret licenses, codes and rules, decide the final requirements, and (where latitude to interpret programs is available) establish what those programs would be (*see* Fig. 3–14).

As indicated in the Japanese Total Quality Management method, the question "why" can never be asked too many times, but there comes a point at which answers no longer add new, meaningful content. An excellent basis process allows detail that

Basis Analysis

Summary: *A basis answers the question "why?" For a maintenance strategy, a basis may answer questions at any decision step in the development process. Why is this EQID key (or not)? Why is this template (generic or applied) needed? Why is this EQID a Master Model? Why are template PM tasks grouped into workscopes as they are? Why are identified parts listed (or not) on a template? Why are failure modes listed and given PM tasks (or not)? This basis map illustrates some key basis justifications useful in RCMtrim.*

Basis Question	Answer/Location	Detailed response
EQID: What makes this component critical?	LTA Spreadsheet Documents selection basis	**Ans:** Functional Impact of failure on system functions as documented by the LTA Risk Exposure (table or spreadsheet)
Generic Template: Why is this generic template needed?	Generic Template Tables Utility & Uniqueness	**Ans:** It provides unique overview of a class of hardware (by design) useful to model for applied templates. Implicitly justified by existing and uniqueness from other templates.
Applied Template: Why is this applied template needed?	Applied Template Tables Application Uniqueness	**Ans:** Identified (by plan) for applied templates based upon master models in the spreadsheet. Justified by uniqueness of master model (plant EQID) from similar equipment. Partly engineering modeling choice.
EQID: What makes this EQID a 'Normal' Model?	Given criticality, it's the best, most representative & recognizable EQID; partly user's choice	**Ans:** It's the most representative and recognizable EQID of its hardware class, which are indistinguishable from each other based on service & environment.
Workscope: Why are specific PM tasks included in the workscope?	Most effective task suitability grouping for work	**Ans:** Partly due to the work nature—like complete disassembly, partly due to failure mode PM task performance intervals, and possibly other factors such as craft involved, routes, or equipment availability.
Parts: Why aren't all the OEM-listed parts identified?	Only DFM* parts are necessary	**Ans:** Only dominant failure mode-causing parts are developed, except to document a failure mode to document analysis .
PM Tasks: Why are displayed PM task strategies selected?	Best tasks, based on judgment & process selection (MSG-3) order	**Ans:** Dominant failure modes are given "applicable and effective" tasks using reference & direct engineering judgment. The assignee may formally provide a selection basis (with the authority table), or leave an implicit basis—the failure mode description, for justification.

* Dominant Failure Mode

Fig. 3–14 Basis Analysis

helps us understand meaning but doesn't require details for their own sake. Requirements for field entries in some software packages make their use so onerous, so burdensome, that users hate them. They add no value.

The beauty of the generic template is in its general nature and application potential. As explicit requirements and their supporting bases come into play, it becomes more important to document and understand broader requirements that may apply and avoid errors of omission in template application. This requires more generic template types and different emphasis. Using the generic template foundation, developing related templates containing rich basis detail speeds work. Copying and customizing similar niche-oriented templates provides an ever more useful timesaving.

Template basis can address the following areas:

- why the template exists
- why parts are included in the part partition
- key part risk information insights
- part failure mechanisms
- key insight on part failure mechanism strategies
- part failure mechanism PM tasks
- alternative PM tasks
- nominal PM task intervals
- basis inputs
- why some basis inputs are more important than others
- workscope organization

If facility management philosophy supports implicit basis development and boilerplate minimization, acknowledging tools and techniques administratively by experts establishes process norms. Clarifying mandatory and optional basis forms or table entries completes basis detail expectation. Details should only be provided where necessary to add value, provide insights, or increase clarity about the PM tasks. Placing an entry just to fill a field should be avoided.

As the workforce ages and replacement personnel take on the work, the basis helps to explain task rationale to avoid relearning past lessons. Processes that check an existing PM basis before initiating a PM work order change assures that past rationale for work is understood before initiating new ones. This is a corporate, strategic maintenance-process step.

As students and instructors engage in an engineering problem or technical analysis, a basis has been expected as part of the answer development. For educational reasons—homework or tests, the basis obtained partial credit when the final answer was incorrect! In all professional cases, the certified engineer must maintain a basis for the development of work with risk implications. From the instructor's viewpoint, the value of the basis to the student was the student's ability to reconstruct the thought process to the correct answer. In work, the basis exercise has been useful in developing better answers over time, or when one is needed as a result of new information. This provides a rationale for why a basis is useful and important. Over time, equipment requirements change. Complex reliability solutions and tasks built from scratch require a great deal of input, thought, coordination, and learning.

It's easy to forget the finer points of any analysis in the volume of work addressed over a few months or years, the interval required to acquire new information that reshapes the PM program. The value of an explicit basis is being able to retain and recover, at a moment's notice, key considerations that selected tasks and intervals apply to high-value equipment failure management. As these change, and they will, it's easy to go back and make course adjustments to keep the PM tasks relevant for new failure information (*see* Fig. 3–15).

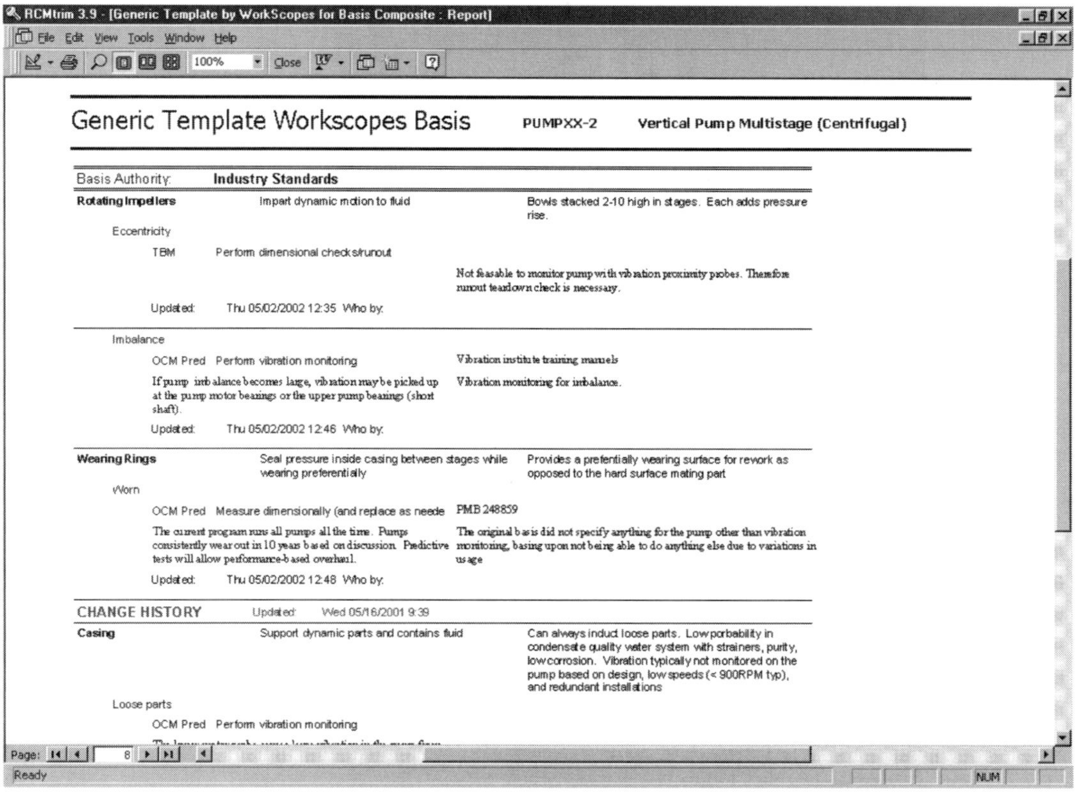

Fig. 3–15 Basis Report for Summary and History

Problems

Standardization vs. customization

Generic template construction can be difficult or easy. While quickly generating a PM program, generic templates don't capture depth and scope embedded in installations. Templates need specific installation details, preserving source material genealogy. This is a tall order—like providing a cookie cutter with ten different finishing techniques all serialized back to the cutter! Template customization must address variation in the following areas:

- dominant failure modes
- aging stress behavior
- risk exposure
- task selection basis
- application conservatism
- equipment application groups

The solution is a *second* template, once removed from the generic template. Generic template application based upon context creates a new entity—the *applied* template. The applied template models *real* equipment, while the generic template models *applied* equipment.

Custom application is a two-step process: select the applicable source template model, then apply appropriate parts, failures, and PM tasks to build the customized applied template. The first step selects the template, the second applies the selection, picking and adjusting dominant failure modes, tuning their PM tasks and intervals for use.

Exhibited failure modes

Useful generic templates identify and develop common failure modes. Characterizing failure details requires failure engineering skill. Consulting engineering textbooks identify engineering failure modes theoretically. At least 20 kinds of anti-friction bearing failure modes can be found in bearing books, for example. Detailed failure mechanisms cause specific bearing failures. Fretting, fatigue, and galling all cause rolling element bearing failure but are of little direct use to the craft. Craft workers would more likely describe failure as wear, contaminants, or inadequate lubrication. Useful failure descriptions mirror shop floor symptoms and descriptions, relating hardware experience for craft use. The craft perform routine work. Useful failure descriptions must be meaningful to the craft performing work.

Contaminated or fatigued bearings fail differently, but the differences will not be evident without detailed metallurgical and lubrication analysis. Failed bearing lubricant and deposit analysis should be sought as part of failure-engineering age exploration in order to develop probable failure mechanisms. *Age exploration* describes the process of defining failure mechanisms completely in aging, symptoms, and related failure mechanism factors so condition-monitoring tasks are consistent with field identification results. Failure descriptions should echo craft terms. Failure-descriptive skills and terms can be learned. Maintaining a database of parts, failure mechanisms, and systems helps develop description skills. Correct complete failure descriptions help pick appropriate tasks, identify correct symptoms, and assign on-condition intervals. These determine service inspection or replacement intervals.

An occasional PM lapse or new emerging failure experience provides an opportunity to adjust PM task content or scheduling based on the new failure analysis learning. Failures observed provide environment, service, and other equipment-condition insights. Inconsequential or unintended failures provide raw materials by which to examine actual failure modes, adjust PM tasks, and validate interval appropriateness. The plant failure record is the best indicator of leading-failure mechanisms. Where preventable, these failures can tailor the vendor guidance on specific equipment applications to provide the best possible PM service. They also identify life-limited, in-service equipment and suggest design changes that will provide the greatest equipment service value over equipment lifetime. These failure-interpretation efforts define and develop an age-exploration process.

Theory vs. practice

Practical failure analysis focuses on the end user—the craft. Academic theory texts on engineering failures have descriptions that aren't useful in the field; they fail to relate terminology to observed, described failures. Unless specific PM task on-condition evidence suggests a new emerging failure, the identified mode should be presumed correct.

A pre-developed PM task based upon fatigue crack analysis need not confirm that each crack specimen found is, in fact, a fatigue crack. This is neither necessary nor useful when expected performance conditions are obtained. Crack failure symptoms consistent with fatigue cracks are adequate indicators of fatigue where this has been pre-identified as the dominant design source of cracking. The presumption could be based upon cracking, crack details, expected life consistency, crack location (high cyclically stressed locations), and other contextual factors.

Reliability engineers should become versed in these subjects by working with the shops that are studying emerging failures.

Enumerating failures

Engineers developing templates can focus on enumerating failures, that is, finding all the possible ways failure *can* occur, rather than the few statistical ways failure *does* occur. Which is more valuable? Neither is harder than the other, but one can't perform the latter without collecting data.

Engineers may focus on engineering development, rather than on useful economic value. Identifying dominant failure mechanisms from craft interviews and statistical failure work orders counters this trend. Other resources available include published engineering failure mode descriptions, vendor literature, and professional codes like the ASME's *Boiler and Pressure Vessel Codes*, regulatory guides, and personal experience. The Internet has a number of professional organizations' sites that are extremely useful as starting points.

Failure enumeration is an organizational-development learning phase in RCM. Reliability engineers experience phases and grow through them. FEMA is another such phase. Interesting upon first discovery, the novelty fades with proficiency. Engineers are fascinated with failure, or they wouldn't be engineers. The key is managing this focus towards a useful end. Without caution, promising RCM-based PM developmental efforts stumble. Failure enumeration should be passed up. Enumeration does not lead to useful PM materials for the craft. Design engineers research failures to develop and commercialize new products. The commercial viability of new hardware rests upon practical utility as well as survivability. Practical hardware failure mechanisms will be limited in commercial applications.

Service intervals

Service intervals depend on part aging, strategy, and risk. Aging depends on fundamental physical phenomena. Physical design, risk design, and installation stresses determine interval selection.

Although mathematics helps identify appropriate intervals, equipment failure processes and actual failure statistics are also helpful. Aging study depends on CMMS/EAMS part tracking. Few company CMMS/EAMS systems in service today provide accurate part history data. New systems are only starting to provide useful part service life data. Parts exhibit *mean life* and age dispersion characteristics. Manufacturing processes introduce product variations responsible for aging-life dispersion; environmental variations in service cause more aging-life variation. Part-service information for most practical purposes doesn't exist. Estimation, guess-timation, and chance help establish equipment lifetimes very uncertainly (*see* chapter 5, Fig. 5–1 and chapter 2, Fig. 2–26).

For example, one case involves the analysis of bus duct cooling fan belt drive failures for two 1350 MWe nuclear units. Observed lifetimes varied from one to six years. Quality belt manufacturers predicted minimum lifetimes of *two* years—the minimum expected life for top-of-the-line V-notch power belts. Belt aging, which determined the PM replacement interval and quality, varied among suppliers. Success depended on procuring and installing quality belts. No supplier would guarantee its belts for even two years, although two vendors predicted their belts would be *fully satisfactory* at four-year replacement intervals.

With belt styles in use ranging from discount auto-part supplied products to world-class, high-quality replacements, belt life certainty could only be assured by procurement. Practice had been to buy non-Q parts on cost. But inexpensive substitutions would not last four years. The risk of low quality was difficult to communicate, discuss, and convey to the buyers. In the end, the plant continued to procure any belt using two-year replacement intervals.

Engineers work under existing policies, and even large production losses can't change cultures. What many know from personal experience is that part quality varies with cost; as the saying goes, you get what you pay for. Quality procurement options aren't valid considerations for a buyer graded on minimizing cost—even assuming they had the skills to assess quality. Yet, fully developed age exploration must include parts history and performance analysis in the overall effectiveness calculation. Knowing critical equipment applications, pointing out low-quality substitutions, and resolutely pursuing procurement reliability helps manage departmental part sub-optimization. Demonstrating a belt causes risk in a high-value (though a non-safety) application also carries weight.

But capability and perception are worlds apart. Statistical quality control uses numerical techniques to evaluate vendors, parts (lots), and materials. Manufacturing ideas may not work in process-facility maintenance. Maintenance processes benefit from manufacturing quality practices. Process maps, statistical examinations of defective products, and Ishikawa fishbone quality diagrams aren't particularly difficult to learn, but old habits die hard. Statistical tools apply to RCM analysis. Failed parts invariably beg the question of their origin.

Third-generation CMMS/EAMS systems track parts. Parts-aging acumen has reliability failure analysis benefits. Failure of parts in service can be plotted versus time, developing the failure distribution curve from raw data. Except for the most expensive equipment failures, part tracking by secondary inference and engineering analysis is not cost-effective to pursue in terms of engineering productivity. Part tracking is expensive; most plants lack the processes to track parts cradle-to-grave, establishing actual lifetimes.

Purchasing from even *two* vendors greatly complicates parts failure study and quality procurement analysis. The lesson for maintenance is to buy from a single qualified quality supplier.

Replacement intervals are based upon risk, redundancy, and part lifetime evidence, shown by a graphical failure *curve knee*. Probability-of-failure distribution curves with lifetime characteristic show a data rise at the end-of-useful-life. Risk determines how *conservatively* to locate the rise (or knee of the curve), for a time-based aging part. Random failure, which exhibits no such distinct rise, must be addressed by design redundancy. Random-failing components require redesign to control critical failure effects (*see* Fig. 3–16).

Direct random failures introduce unacceptably high safety or operational risk, high cost, or both. Low-risk, cost-based failures one can accept. Redundancy is a simple random-failure risk control strategy. Redundancy can extend to any component, regardless of failure nature. As electronics sensor and circuit costs fall, design redundancy as a control strategy is economically more compelling than ever.

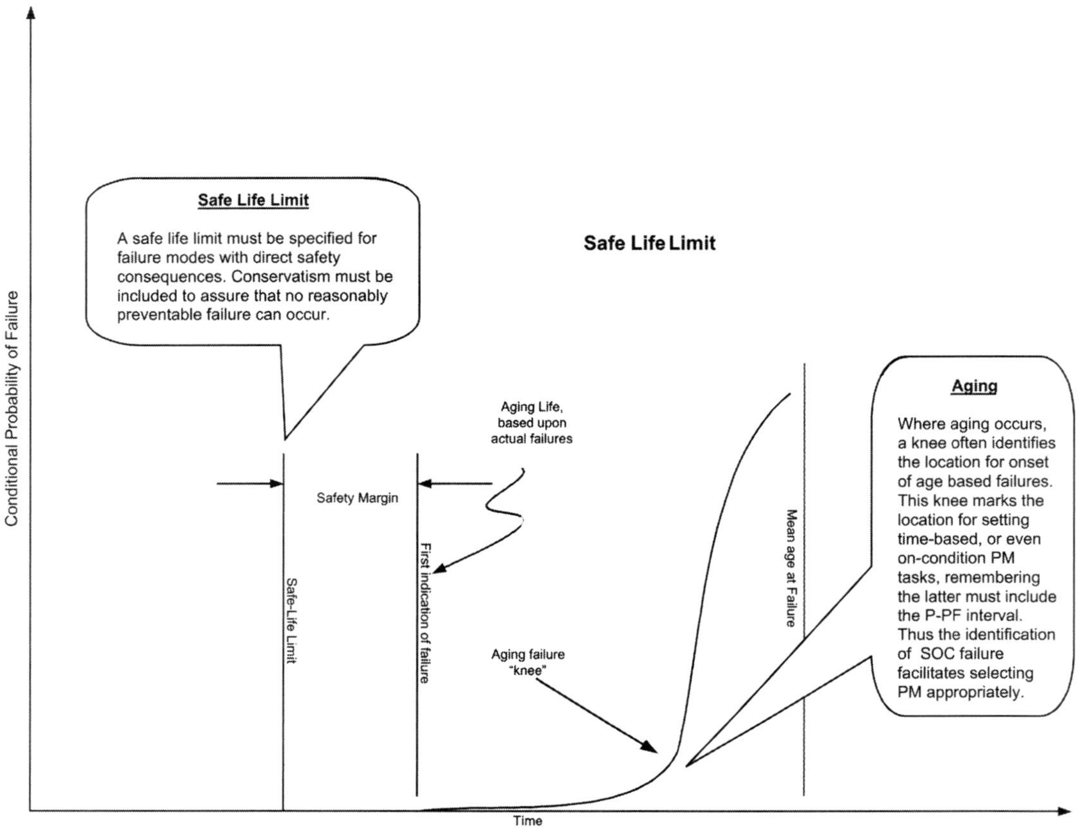

Fig. 3–16 Failure Knee for Life-Limited Failure

Failures are seen as concrete, deterministic events. Yet, viewing statistical data using probability concepts, this failure behavior is idealized. RCM addresses single-failures. Failure data validates the existence of single-failure modes. Reviewing failure data plant-to-plant establishes benchmarks. In cases where multiple failures are present, Weibull analysis technique reveals multiple failure patterns. Weibull graphs extract single-failure modes from complex failure data. Plotted on Weibull paper, multiple failure mechanisms exhibit slope change.

Failures themselves are best viewed in low-tech media. Simple data presentation and written text interpretation can effectively present failure modes. To address a failure mode, one must first identify the mode. This can be tough with many vaguely worded work order failure descriptions.

Workscopes

A workscope provides an efficient way to group separate failure-preventing tasks under one common work package for efficient tagout and work performance. The workscope differs from a PM task. Striving to minimize work trips, workscopes group tasks based upon skill, equipment tagouts, and craft availability. Tasks address unique, engineering failure mechanisms. Workscopes address scheduling and planning utility. From their failure modes, tasks can be developed independently from workscopes. This allows the engineering work basis to evolve separately and independently from work orders. Work order tasks organize the scope necessary to complete the work, which comes from the approved work tasks (*see* Fig. 3–17).

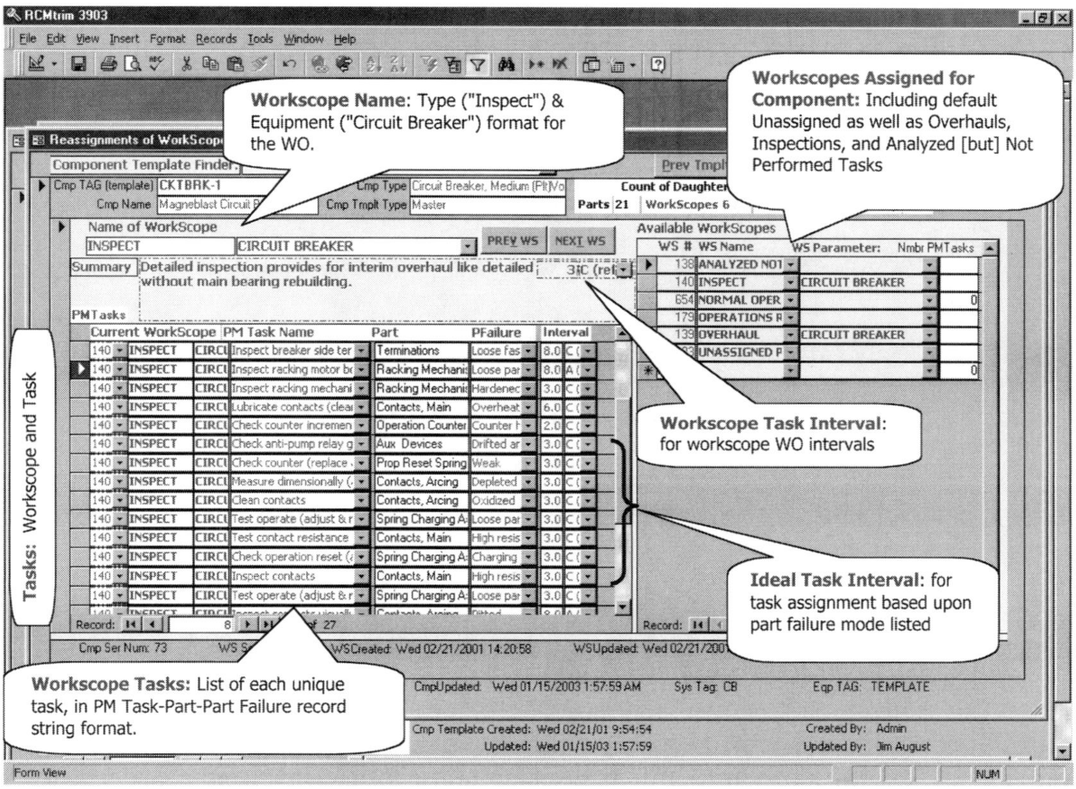

Fig. 3–17 Workscope Task Blocking

Individual tasks tasks must be scheduled considering failure characteristic (mean time between failure (MTBF), Weibull parameters β and μ. These determine the failure mode aging and life characteristics) and failure risks themselves. Failures affecting safety have safe-life limits for aging failures with non-redundant backup. All failure modes have PM intervals and a failure control strategy—time-based (hard-time) or on-condition monitoring tasks. On-condition task intervals must be selected considering potential failure-to-failure (P-F or PF) intervals. Individual task performance intervals must reflect failure time-dependence characteristics, PM strategy, and risk based upon that failure mechanism. When individual tasks are rolled-up into workscopes for common work order performance, adjustment and compromise is needed (*see* Fig. 3–18).

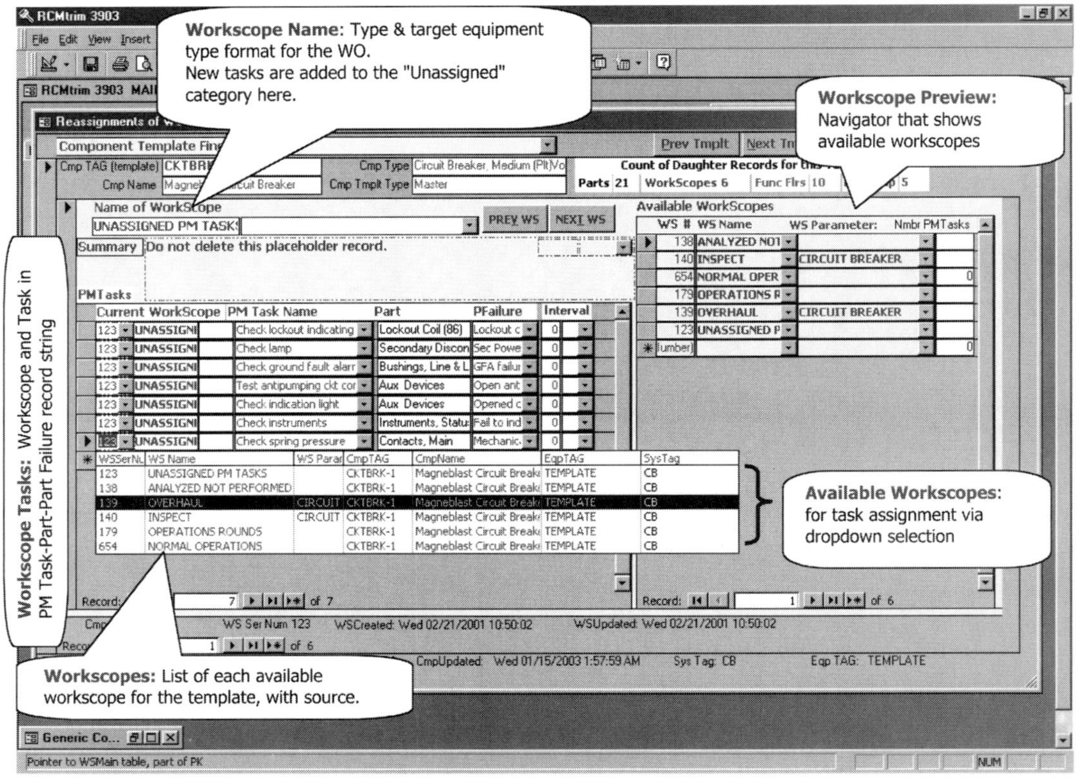

Fig. 3–18 Workscope Task Re-Assignment

Workscopes optimize work, forming individual tasks into blocked groups. Blocking—scheduling groups of tasks to be worked together based upon fit—scheduling appropriateness, tagout, other work interference, skills, intervals, etc., all require work knowledge and craft worker abilities. In theory an applied workscope interval can't be longer than the shortest PM task interval of the combined group. Practically, the hard limit applies rigorously for safety-based failure mechanisms but can be flexibly applied on production or cost based mechanisms, based on risk. Operational limits provide a bit more latitude than safety, depending on the operational risk; cost can be adjusted considerably keeping cost alone in mind.

Workscopes also allow the roll-up of the time and cost information to perform work. Time can be delineated into the time spent addressing various task failure mechanisms in the WO—associated time necessary to travel to and from the job, pick up tools, get parts, and so forth. This provides a natural breakdown for estimating work time and separating work trip time (which depends on the travel to and from the work site). Specific task performance times are easy for craft to identify; broader overhead and work coordination times are easier to develop by comparing other job experience. Planners estimate worker tasks well. Jointly, task times total into work-scopes for costing WO work by equipment, equipment category, and other costing criteria.

Historically, maintenance departments don't closely follow WO costs. As production becomes more competitive, the need to understand maintenance costs increases. Providing the work constituents that drive scheduled maintenance work order costs based upon value (failure prevention) starts the process of identifying work value. Time estimating is difficult for maintenance organizations; estimating avoided costs from performing scheduled maintenance is even harder. However, cost-benefit calculations for each PM task, based on the avoided-failure cost, allow development of a cost/benefit PM ratio. Engineers with cost insight must perform these estimates.

There is no way to value public or employee safety. Tasks that directly prevent safety failures are simply *performed.* Operational (production) loss-avoiding tasks historically outweigh many times over the most expensive maintenance failure costs. Operational tasks can be approved with few cost benefit worries. Direct production losses are avoided by performing the tasks; those losses outweigh tasks by at least a factor of 10. Cost-based failure is the sole remaining cost-benefit calculation category. Performing benchmark cost comparison cases help quickly establish any comparable task's cost basis. Benchmark cases develop naturally once an organization decides to quantify its scheduled maintenance program in cost/benefits terms. It's easy to reference and used similar analysis to develop enveloping cost-benchmark comparisons.

When an expensive compressor crashes (for example a $2 million fully redundant soot-blowing air compressor) rebuilding costs contrast against the cost of a $200,000 biennial finishing-stage compressor-wheel overhaul. It's easy to quantify PM avoidance in cost benefits. Here the ratio is roughly $200,000/$2 million, or 1/10.

There are three rough cost/benefit ranges: 1/1–1/10 (low), 1/10–1/500 (mid range), and 1/500–>> (high). For various consequences, one can bracket cost-based failures with these three ranges. They suggest the level of effort to place on benefit/cost PM measures. Marginal improvements—ratios of one-third (<<1/3) or less—should be considered only with other work. If new tasks can be incorporated within an existing workscope, they're acceptable; otherwise, they would require a new WO and probably don't merit performance. Workscopes that increase the total number of plant PM WOs should be viewed skeptically.

Last, the total cost and benefit of the work itself warrants consideration. A low cost/benefit may work when the absolute benefit is more than $100,000. At values less than this, it warrants caution. Unless the plant has a cracker-jack maintenance performance efficiency and corresponding low costs, it's probably a cost-based loser!

4

Applied Templates

Strategy

Custom uniformity

Applied templates reflect the requirement to customize generic template models to reflect context based upon risk. Templates found in CMMS/EAMS systems of the 1980s and 1990s were effective tools to extend and standardize PM programs. Models used at that time failed to reflect actual field equipment use and risk and to tie tasks back to failures for quick update based upon field experience (*see* Fig. 4–1).

The applied template bridges standardization and customization to address the combined need to develop pre-defined, expert standards while customizing to exact equipment requirements (*see* Fig. 4–2).

Risk observations

Very detailed equipment PM models can be crafted to accommodate different component designs. Applying these models directly to different components with text documents or spreadsheets requires static application. This is the PM optimization (PMO) approach. The available technology, reflected in documents and spreadsheets, cannot apply complex equipment features based upon contexts because text and spreadsheet formats are limited to flat-file representation. Users cannot depict specific dominant failure modes or custom usage contexts without editing text. *One size fits all* defines the PMO process and limits application to a few standard equipment categories.

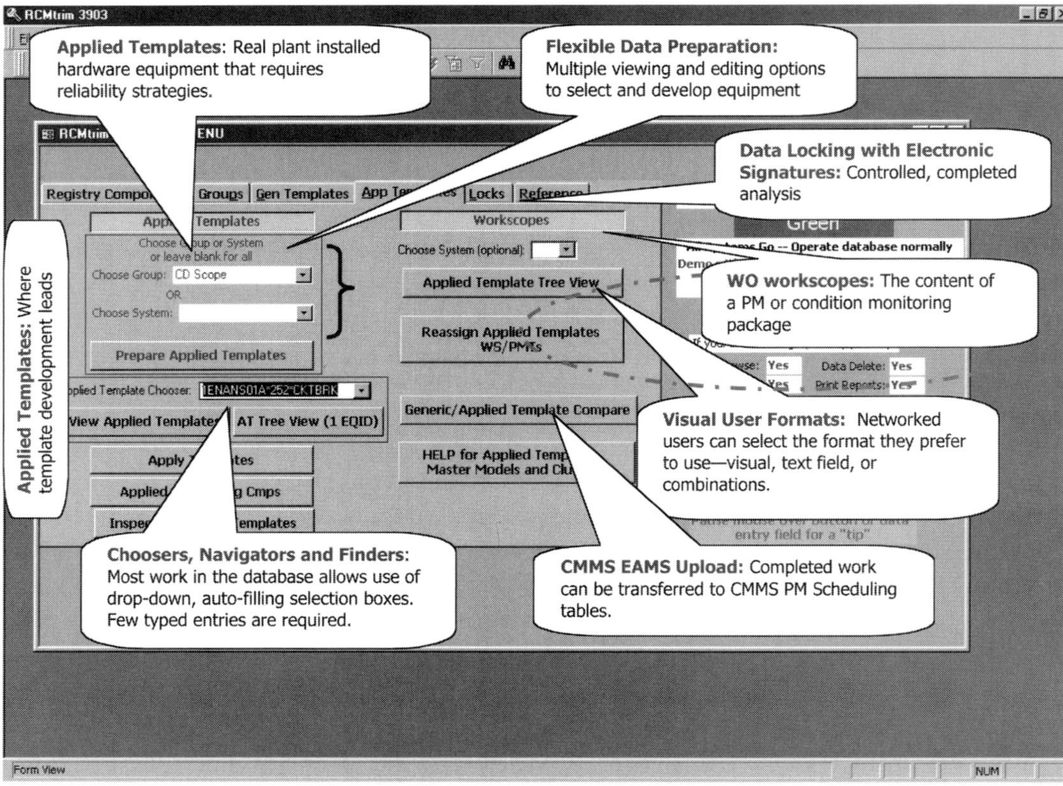

Fig. 4–1 Applied Template Controls

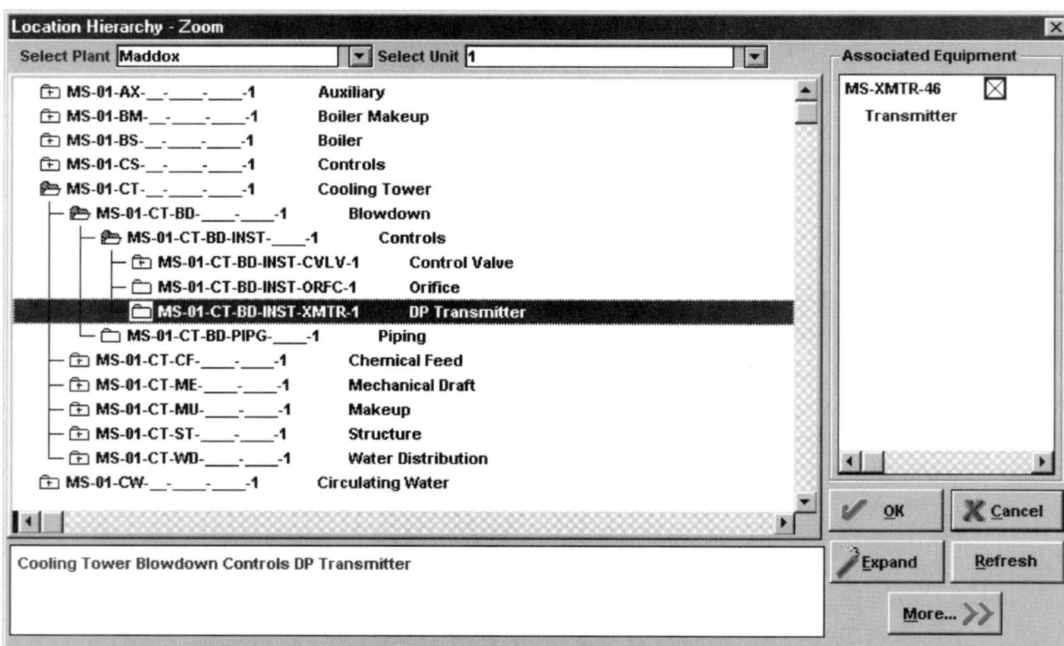

Fig. 4–2 CMMS Plant Equipment Hierarchy

PMO models the most complex versions of equipment with the most conservative intervals and applies these uniformly to all similar component types. Differentiation is limited or non-existent; contextual depth of application based upon risk and service factors can't be accommodated due to the number of independent variables. Common-basis text applies to all template variants of one component type; therefore, special application requirements intermingle with the generic ones. The result is complex programs that include non-applicable equipment, and application errors abound. The means to apply tasks that are appropriate for each component tag separately just aren't available.

In the early years of nuclear plant administration, nuclear generation motor-operated valve programs, for example, upgraded all equipment to NRC generic letter (GL) 89-10 standards even though many valves were balance-of-plant, non-nuclear applications. The justification was the desire to avoid the complexities and risk of running two different programs increasing the potential for errors. All calibrations met the most restrictive safety-related specifications because means to track calibrations separately weren't available. Since craft might confuse separate program requirements as they interpreted work orders, it was safer (based upon compliance) to elevate all programs to the highest possible standard. Many iterations of this process led to ultraconservative programs applied to lower tier, non-critical equipment—even non-applicable run-to-failure equipment such as low risk secondary system status indication loops. EPRI's effort in the 1980s to introduce RCM to help control nuclear PM costs sought to address extremely conservative practices. Reducing established programs based on rational re-analysis proved difficult, however.

A solution to this dilemma would improve capabilities to trace requirements back to functions, failures, first principles, and failure mechanism analysis component-by-component, failure-by-failure. At this detail level, assurances could be provided that would allow mangers to comfortably know that the analysis basis was correct, and the PM task work (e.g., workscope) could be controlled (*see* Fig. 4–3).

A goal for template application is to create an exact auditable trail that succinctly and clearly relates component PM requirements to the generic template through the applied template. Achieving this goal provides workscope implementation in a seamless data environment. The process must be so simple that it meets the needs of PM power users—like nuclear plant and aerospace crafts, as well as those with more modest documentation needs—like fossil plants and refineries who view the primary goal of a PM basis as managing costs. (More industries face increasingly prescriptive PM programs today, as federal, state, and local regulations engage in more directive roles addressing workplace, community, and public environmental concerns.)

Application requirements

Equipment information—contextual risk expressed as "how bad does it hurt?" and statistical likelihood—expressed as "how likely is failure?" establishes risk and ranks failure mechanisms. These dominant failure modes identify PM tasks that matter. For broad classes of equipment, the utility of PM templates is to identify common design

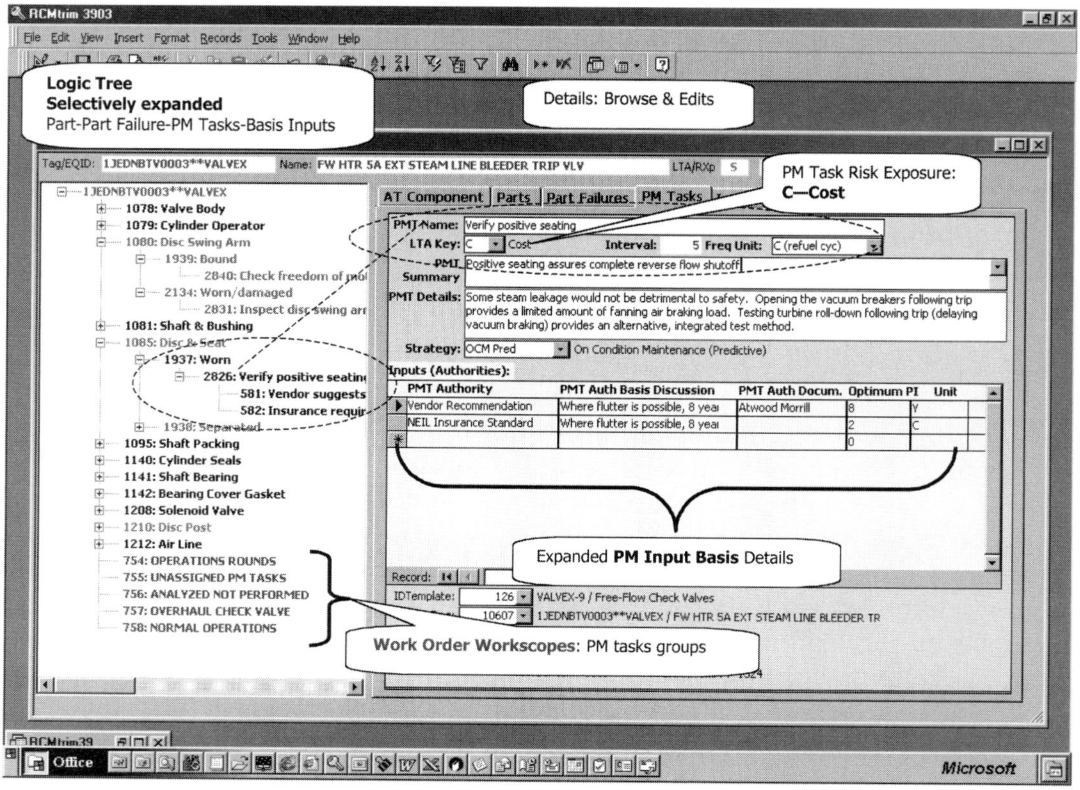

Fig. 4–3 Selectively Applied Parts, Failures, and Tasks

characteristics that provide valid risk groups. Just as actuaries claim that statistical success depends on grouping, successful template application consistently groups hardware by design and manufacturer. Failure modes, mechanisms, and preventive measures then can be selected quickly for reuse. Standardization occurs automatically (*see* Fig. 4–4).

Summarizing, functions

- arise from design
- introduce failure modes that depend on intended component functions
- introduce failure mechanisms that explain how functions can be lost based on design and physics

Applying templates (*see* Fig. 4–5) quickly requires the engineer to do the following:

- select an appropriate model
- find critical functions
- apply part failure/PM tasks based upon functions in context

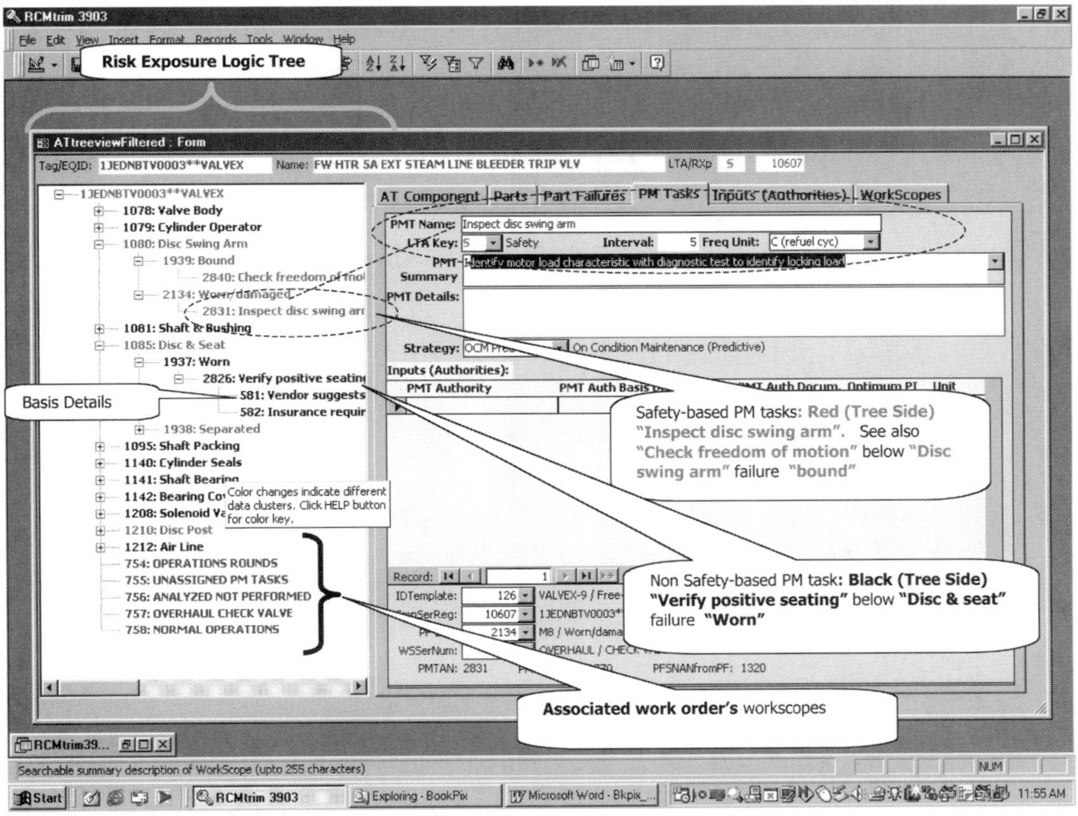

Fig. 4–4 Applied Template Failure Mode Risk

This action quickly generates failure mechanisms and their associated PM tasks. Appropriate, applicable, and effective PM tasks vary depending on context. Knowledge of the failure risk exposure combined with the failure mode identity suggests the PM tasks to select. In special cases, safety for example, the risk exposure classification allows PM task combinations to ensure a failure mechanism will be controlled to an acceptable risk level. Differentiating part failure risk exposure allows the selective consideration of tasks. This allows for progressively selecting failures and their mitigation PM tasks on their risk merit—probability of occurrence, consequences of occurrence, and applicability/effectiveness test criteria. Progressively selecting failures for PM tasks by risk provides these benefits:

- focuses resources where they add the most value
- selects preventable task opportunities within focus
- tunes and times tasks to specific equipment context

Selecting the best PM task mitigation and maintenance opportunities requires customization (*see* Fig. 4–6).

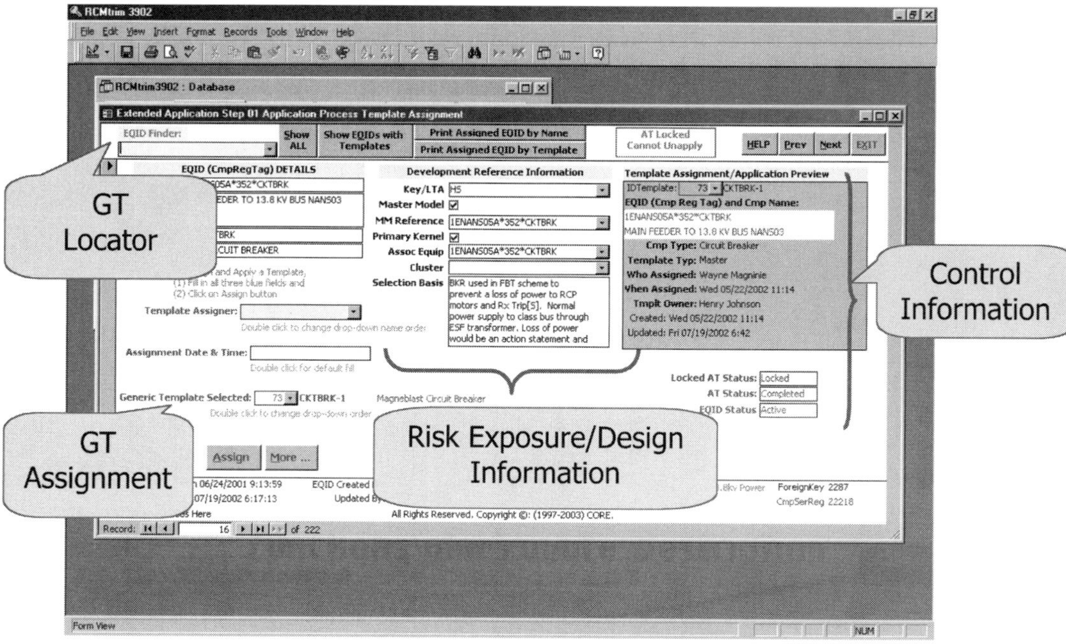

Fig. 4–5 Application Details

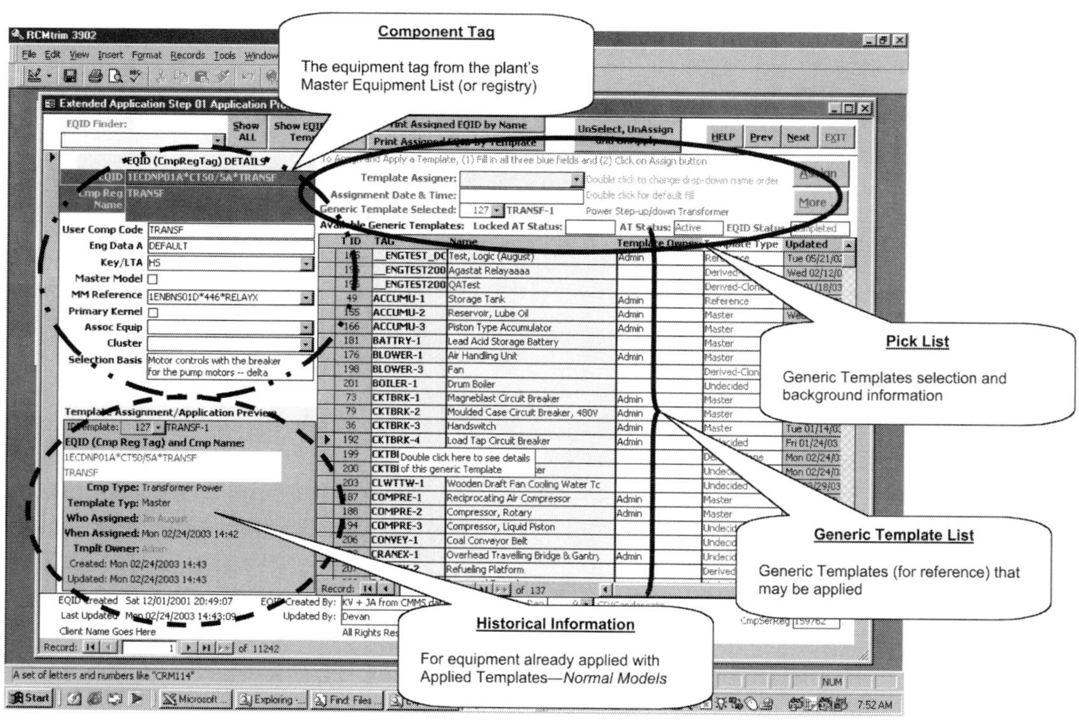

Fig. 4–6 Selection of the Generic Template (to Apply)

Template application and customization

Template application requires knowledge of equipment service, environment, and risk. Together, these define equipment context. Risk arises from system equipment criticality—SOC and the hidden or evident nature of equipment failure. Service and environment come from equipment context. Knowing risk exposure factors, reviewers can select appropriate failure modes based on associated risk factors provided with an equipment template (*see* Fig. 4–7).

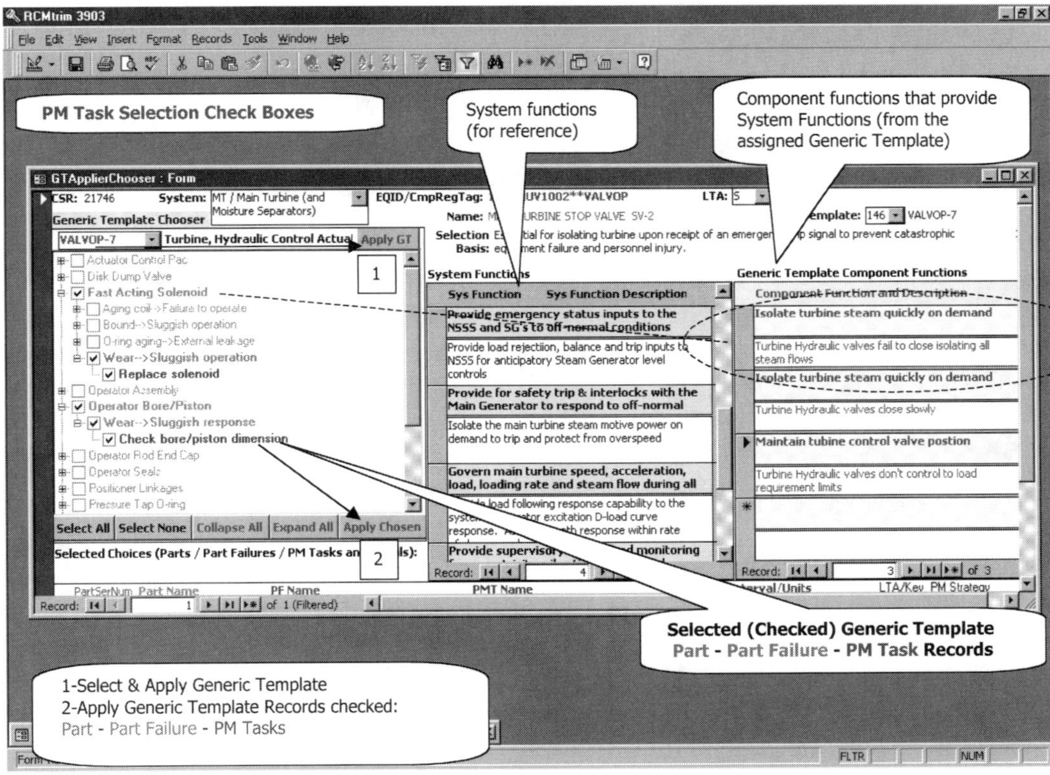

Fig. 4–7 Template Application Steps

Equipment possessing safety implications but low service (most stand-by equipment) should receive the opposite of aging measurement emphasis, usually performance testing. Conversely, functional testing is pointless for equipment already in service. Predominant application characteristics are easy to select from pre-developed templates that list failure mode by parts for typical installations. Template efficiency in these cases is great!

Standards call for template use when developing equipment PM tasks. Templates require identification of common equipment classes with their parts, failure modes, and most effective representative PM activity (*see* Fig. 4–8).

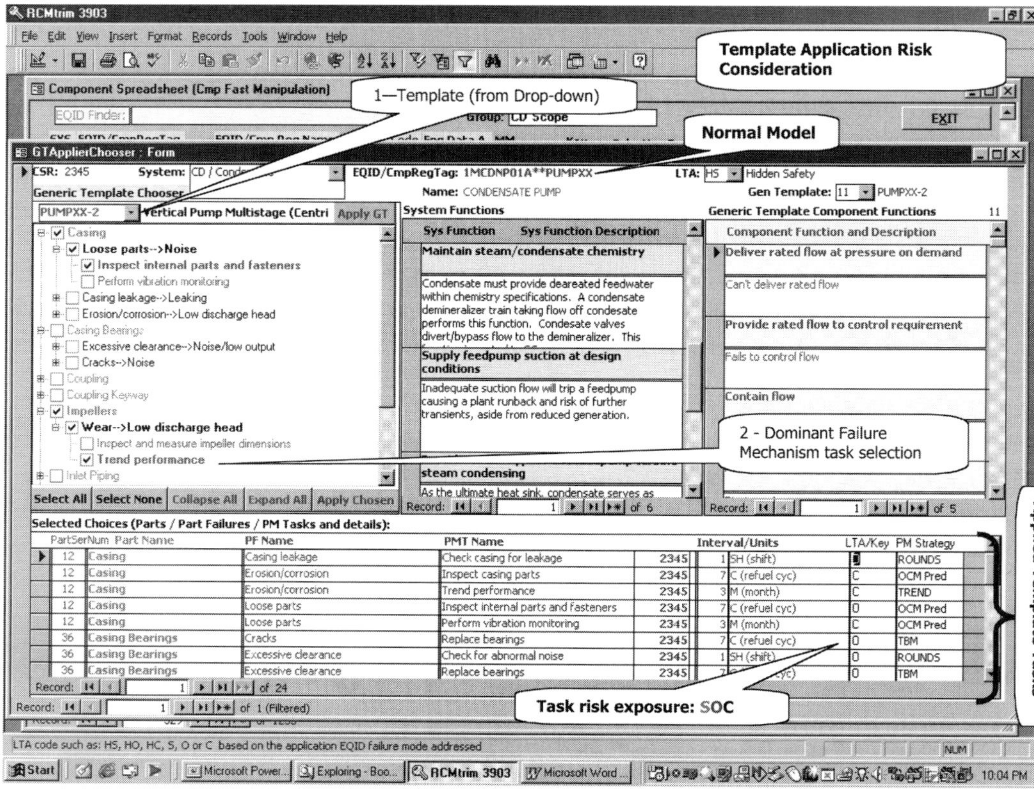

Fig. 4–8 Template Application Risk Exposure

In developing applied templates, one typically limits failure modes and tasks to those already identified on the generic template source. Ideally, analysts can add new ones to the applied template on the fly. A template-based RCM process must provide for template application standardization that retains flexibility in use. It should provide the ability to regenerate new work while enabling analysts to capture and exploit common experience quickly. Capturing and sharing "tribal knowledge" from an analysis/review perspective is another RCM objective. This centers on understanding failure, required performance, and the causes for performance in the PM task basis.

Case in point: At a chemical fabrication facility, a mechanic had devised a simple, non-technical test for locating failed bearings in a long drive assembly. (Bearings were assigned a run-until-failure risk level.) Upon detection of a failed bearing, the line was taken down for a brief period to replace it. Everything worked well, providing an operator correctly located and tagged the failed bearing for the shift maintenance crew before bearing seizure actually forced the line down. One mechanic devised a simple test that identified bearings in failure (PF) with a 1–2 day lead-time. Compared to other lines, his line ran better, but no one knew why.

Selecting relevant failure modes

Selecting failure modes requires identifying failures likely to develop in a given environment. Production facilities with an operating history can review past failures; however, facility age and equipment aging depending on the failure mechanism, make task selection based upon single site experience risky. There are no leading age samples, such as those used to assess fleet aircraft models. Balancing site history with industry experience, manufacturer knowledge and peer experience reduces failure surprises. Systematic acquisition and assessment of opportunity samples further reduces risk of unforeseen emerging failure modes (*see* Fig. 4–9).

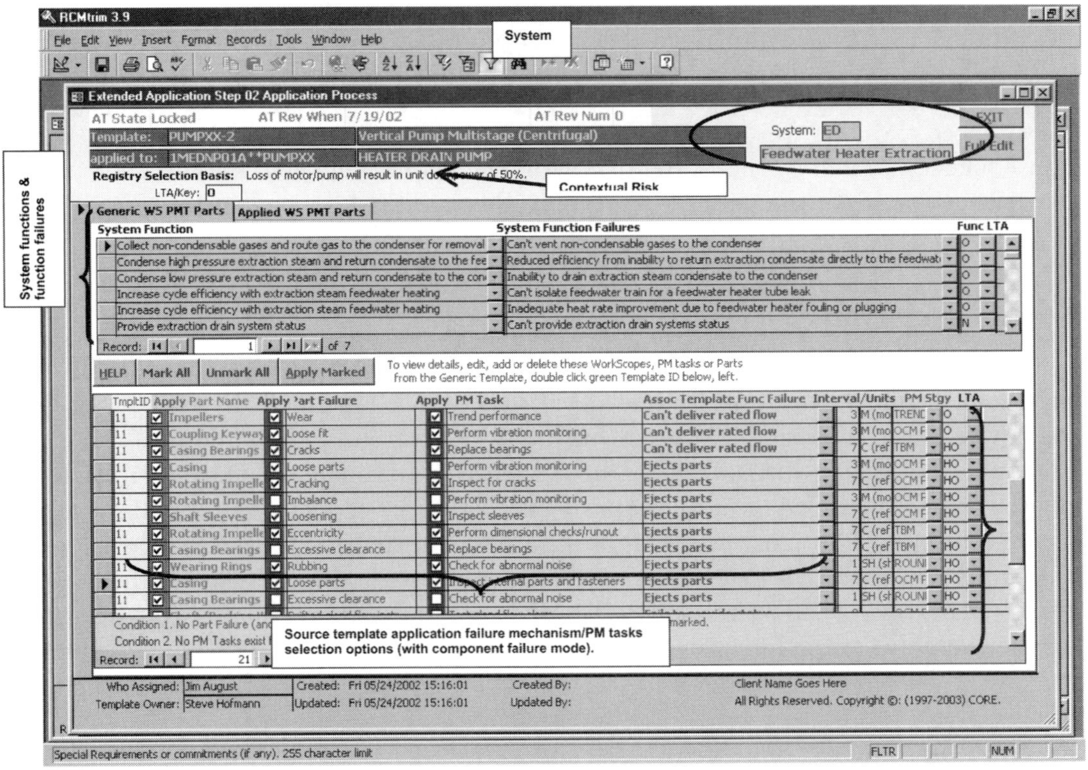

Fig. 4–9 Dominant Failure Mode Expression on Component Function

People *in the equipment* (i.e., working with their hands on the hardware) gain insights into failure modes as they see them develop. Their insights provide useful suggestions for causes of and solutions for further problems.

Suppliers identify high-risk failure modes that affect safety, explicitly providing for their control. Safety alarm trips and interlocks identify otherwise hidden failures. High-risk hidden failure warning logic schemes is also incorporated into design.

Explicit safe-life part replacements are identified in vendor literature, so their review is mandatory for safety. Codes, rules, and regulations express safety requirements. Organizations operating equipment covered by codes have experts familiar with codes that relate to their equipment. Facility licenses often identify high-risk failures and control requirements.

Licenses affect safety, so license noncompliance has utmost risk consequences aside from obvious legal implications. Licenses are sometimes misinterpreted, misunderstood, or not applied as intended. Some plants view environmental requirements with a jaundiced eye that obscures legitimate environmental risk-control concerns. This traditional perspective is vanishing due to the public's support for environmental issues. Attitudes influence risk. Dust suppression and methane alarm systems in coal blending facilities provide case histories of inoperable equipment being ignored. At that point, design intent for these features has been compromised.

Experience enables people to identify risk. Today's work force is aging, and experienced workers are retiring. Younger workers must relearn the industrial risk consequences known by their veteran coworkers. Capturing tribal knowledge from the experienced workforce justifies long-term RCM-based risk management.

Adjusting intervals

Where failure costs alone matter, stretching part-failure inspections and replacement intervals offers economic opportunity. The systematic assessment of parts in service through periodic removal, inspection, and assessment requires a plan to anticipate part-aging mechanisms. Under such a plan, aging estimates can easily be revised, based on experience, to establish a realistic part-service lifetime. This is classic age exploration. To encourage appropriate age extension, failure risk exposure *traceability* is useful to determine component and part failure mechanism by SOC. Once users know a failure mode has clear cost and/or non-critical consequences, they can confidently extend intervals.

Crafting PM

PM development is an art. Standards specify what and how to present in scheduled maintenance plan content. Military, engineering (SAE JA-1011), and industry (ATA MSG-3) PM development standards are available but not widely used. Widespread PM standard adoption could restrain other trends like formal maintenance program inclusion under statute laws. Well-intentioned lawmakers generate explicit legal requirements influenced by public pressure. In addition to mandating maintenance performance, these laws add legal complexity to the production environment. Statute law is notoriously inflexible in a dynamic world.

Lawyers are not engineers. Industry should develop standards and self-regulate, led by the most capable people with the qualifications and abilities to create such standards. Lawmakers should certify industry-engineering standards by code.

Improved maintenance standards and their interpretation would also improve technical compliance. This melding of legal and industry objectives would have direct safety, economic, and other intangible benefits. Standing committees within the ASME, IEEE, and ANS continue working on industry maintenance requirements and standards to support code and other legal requirements.

Vendor dilemmas

The vendors' dilemma is similar to that of other trade or oversight groups: They supply equipment but frequently aren't aware of equipment application ranges. Ordinary pump motors are installed in wet, submersible environments; low-temperature elastomers see high temperature usage; low-strength bolts see high load, cycling application. Design stretch occurs—sometimes intentionally, sometimes not. Run-to-failure is acceptable, provided the application is based on cost, rather than on safety. The designer takes full responsibility for design risks. A design engineer must accept and manage design risk, which includes equipment selection. Vendors can't anticipate designer creativity or every installation technique. They can only specify intended use and expect (or hope) that designers exercise diligence and common sense. (*see* Fig. 4–10)

Creative, outside-the-box uses for equipment may originate with operating and maintenance users who want to fix failed equipment quickly to restore failed equipment to service. These users may not be capable of evaluating risks. Some jury-rigged worker arrangements over the years have both amazed and impressed plant engineers, challenging their engineering mentality. One extreme example involved a half-mile coal belt 7kV breaker load center that kept tripping on belt start up. The operator solved the frequent tripping problem with a plastic spoon in the belt drive motor's breaker trip relay contacts! The intended solution—shoveling the coal off the belt to start it unloaded—was not popular nor even seriously considered.

Application basis

Formal basis explicitly answers why a PM task is performed. An implicit basis comes from the relationship of part-failure to PM task definition. The implicit basis reflects inherent equipment part-failure relationships and all activities addressing the failure mechanism.

Component-applied template dominant failure modes anchor applied PM tasks. Applied part-failure PM task records characterize the component's dominant failures and their PM mitigation tasks. Both generic- and applied-template PM bases input entries directly support PM tasks with explicit justification statements. They generate an *explicit* basis, while part-failure mechanisms with corresponding PM strategy create an *implied* basis (*see* Fig. 4–11).

PM inputs into the PM bases—creditable insights and recommendations from qualified authorities providing basis for PM requirements—detail explicit recommendations, requirements, or other considerations and the way they address PM tasks

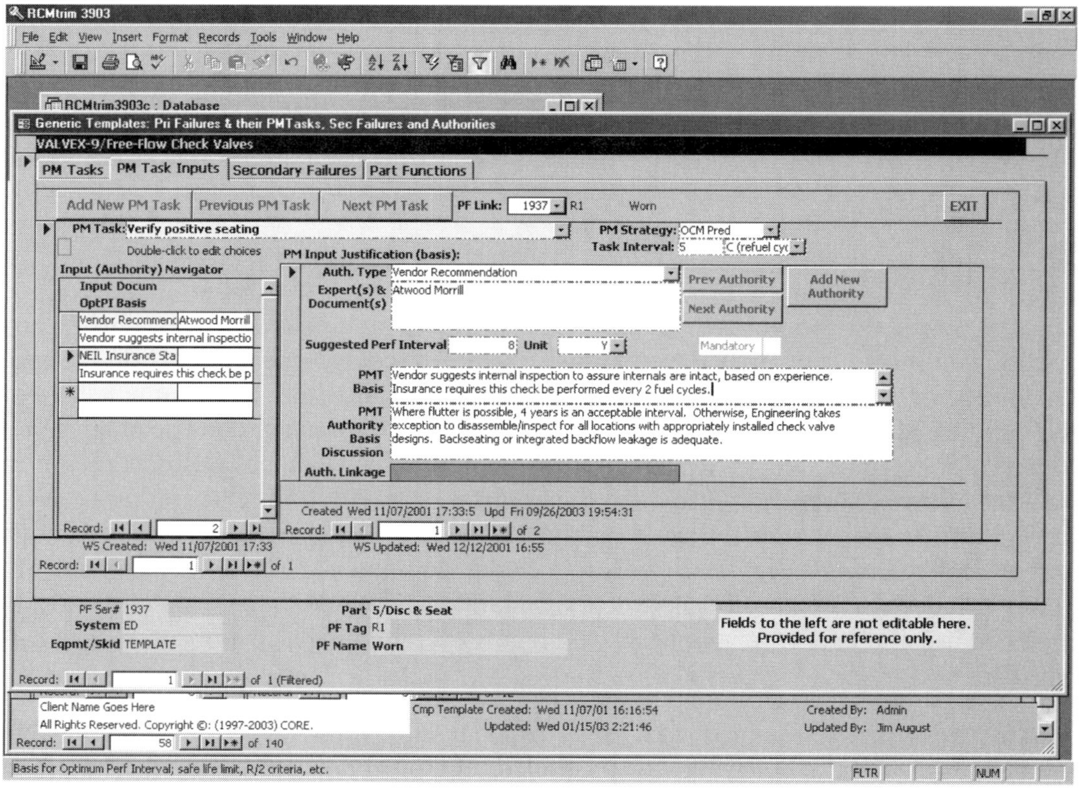

Fig. 4–10 Making an Exception to Vendor Recommended PM

and failures. While an applied-template explicit basis need not be constructed in cases where engineers, standards bodies, codes, or statute law have addressed complex failures, these explicit requirements should be documented. PM input references (a.k.a., PM authority references) perform this role. Explicit operational requirements associated with preventable failures have direct scheduled maintenance requirements.

PM input task association makes the connection between mandatory PM or surveillance requirements and associated code, license, or regulation explicit. Interpretive requirements provided by experts can also provide opinions at this level. Any PM or any PM task attribute change such as task content, scheduling interval, or performance limits should be justified or as changes in the PM input fields.

An explicit basis value comes from documented compliance to code, statutory law, and other requirements and from connections made between these requirements and their intended targets. These connections funnel upward, component by component, through failure modes to the component system's supported function.

PM requirements that address required tasks affecting the public health and safety or that relate to matters of public interest due to their perceived impact on societal goals have the force of law. (Sometimes the broader challenge is interpreting which failure mode the regulations have targeted by force of law.) Societal goals include

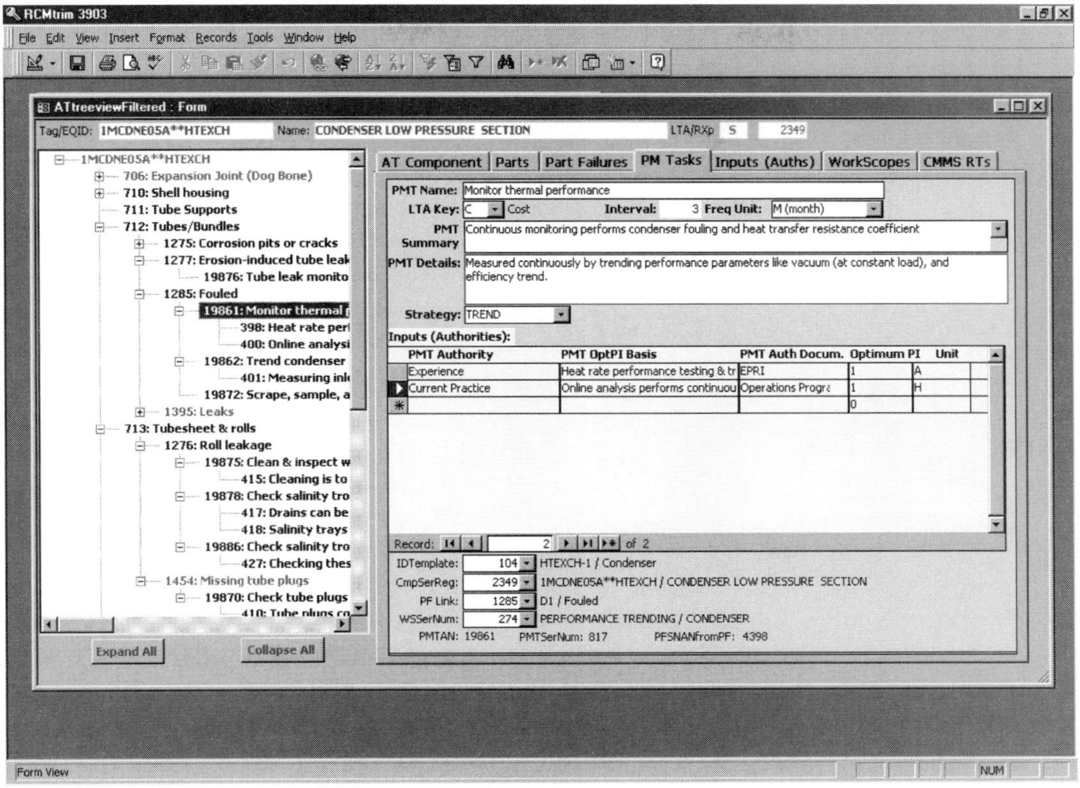

Fig. 4–11 Applied Template Task Basis

environmental and workplace safety matters. Without judging merit, where these requirements are in effect, they deserve structural compliance. *Operationalizing* activities that address these requirements cost-effectively is good business policy. Achieving compliance in a simple, cost-effective manner is the objective. Documenting mandatory basis in a PM input basis, and providing a relational PM-failure/part-component WO helps demonstrate compliance as part of the facility's routine work practices.

Intrinsic basis

An applied template's intrinsic basis depends upon specific equipment installation. Failure statistics or operating observations reveal a failure profile. Equipment operating in moist, high temperature environments exhibits problems different from those in cool, climate-controlled areas. The former scenario characterizes a cooling tower; the latter, a nuclear standby safety-injection pump room.

An *intrinsic* basis helps develop the *extrinsic* basis. An applied template captures specific installation context requirements. It represents real, plant-installed equipment in an actual plant. The applied template interprets theoretical problems through a tinted glass of one component in one plant (*see* Fig. 4–12).

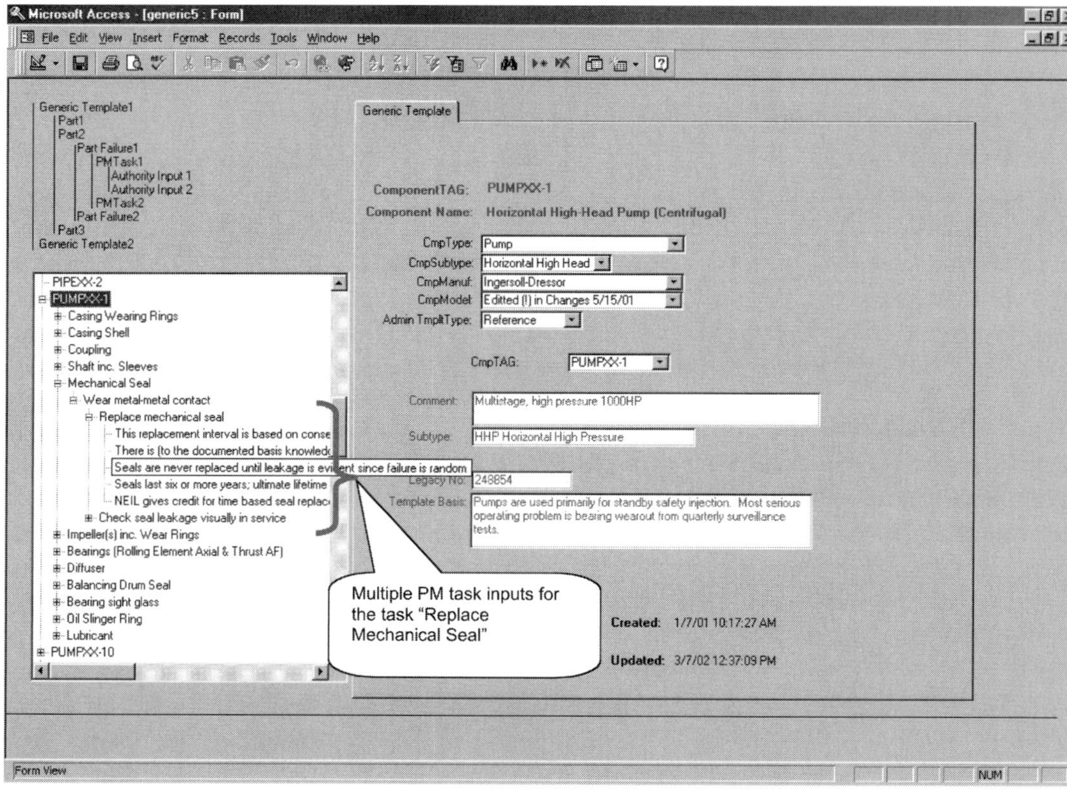

Fig. 4–12 Applied Template Intrinsic Basis

Equipment in service tells a story; each component has a history. The history reveals better than anything else the equipment application's stresses and performance in a specific situation compared to nominal or ideal performance identified in a vendor's O&M manual. Equipment must have been in service for a long enough period to develop a history, of course. For many plants, this period is several years. For plants with very protected environments and low equipment stresses—nuclear plants, for instance—the period may be 10 years or more. The time needed to develop representative emerging-failure samples statistically depends on the aging characteristics that cause failures.

Many equipment parts don't exhibit aging until plants enter mid-life, perhaps 20 years of age. At this point, elastomers harden, lubrication dries, switchgear lube hardens and heat exchangers foul, and electronics/electric equipment shows dielectric resistance changes. Semiconductor breakdown and insulation resistance deterioration show up as increased partial discharge; voltage ripples and logic errors from circuit redundancy loss are embedded deep in microprocessor designs. Power supplies fail, capacitors leak, and previously failure-free parts exhibit rising failure rates.

Intermediate aging-failures are best identified from industry and manufacturer failure databases or by benchmarking similar equipment in older installations. Industry experience can project the plant into any future period to project statistical failure numbers and obtain *leading age* performance samples. However, lowered stresses in some industries inherently block development of the very failure data needed to project future problems.

This is the dilemma: To identify future failure patterns, one needs leading part- and component-age candidates to assess performance expectations. Yet, for safety-influencing failure modes, failure isn't an option. Instead, test applications must be used with accelerated aging to *preview* future performance. Accelerated aging methods are well documented, and manufacturers carry considerable amounts of aging data from product development that indicate how parts perform under *anticipated* service conditions.

Developing failure statistics from *opportunity samples* happens informally as plant workers maintain equipment in the plant. Capturing this information to develop future age performance estimates requires commitment from the maintenance shop and reliability engineering. Database-aggregated information is also necessary to relate failure modes to materials, location, service, and use. Early CMMS/EAMS systems promised this aging data; practically, training, committed effort, and elegant database design is needed to make information develop from this data.

Sometimes low-tech is the best solution. The old calibration cards files, which tracked loop calibrations, gave way to CMMS/EAMS computer technology. CMMS/EAMS systems lacked user-friendly interfaces, and facilities lost an effective, low-tech way to track operating history for instruments when card systems were abandoned. Operators making *round notes* on equipment cards were another effective technique for recording equipment history. Systems that collect operator round notes or other periodic servicing comments and input them into computers require commitment and understanding of the workers expected to do input data. Too often, plant administration asks for more than back shops can provide. The craft typically has the last say in matters like this.

Systems that documented equipment failure offered CMMS software add-ons for vintage early-1980s CMMS systems. Some failed because they lacked user insight needed to reduce worker observations to WO entries simply and easily. Workers won't support an inordinate amount of extra effort. Several such systems required *failure cause* field entries to close WOs. At first completed WOs stayed open in the CMMS/EAMS, then they were completed superficially as workers found computer completion codes that *worked*. (*Worked* meant that the CMMS/EAMS software accepted the code to closeout the WO.)

CMMS/EAMS systems' failure histories are always inadequate, even where failures occur frequently. Fossil coal-generating plants in service more than 20 years generate several thousand failures per unit per year. The associated WOs represent real, failing equipment and provide a material source to develop Pareto component failure distributions. Statistics are not an end in themselves; they simply show where the next stage should focus. But Pareto failure data distributions provide insight about type, frequency, and consequences of installed equipment failures.

Development of failure modes and mechanisms follows the identification of failed parts. Operator, craft, and technical expert interviews help engineers focus on aging behavior, failure mechanisms, and those insights relevant to fully develop failure modes at the part-failure mechanism level. Parts fail from physical stresses—voltage spikes, stress cycling, erosion, thermal aging, and a host of other problems. Therefore reliability engineers should be familiar with basic failure processes at a working level. Exact crack causes could be fatigue, material, or corrosion among a host of other potential engineering mechanisms.

The good failure engineer is like a family doctor. He has the insight to know probable causes are based upon failure context and the wisdom to know when to call in the experts. In the interest of economics and time, failure engineers should be familiar with most common failure symptoms and contexts. They don't request metallographic replication analysis of every high-strength steel crack in stressed, cycling applications to confirm suspicions of fatigue cracking. However they might require expert support for cracking analysis in stainless steel boiler superheater tubes, where tubes were known to have been washed down with chloride in an unusual event and where unusual cracking problem developed prematurely, inconsistent with the normal fatigue aging life of the tubes. Reliability engineers are Jacks-of-all-Trades, possessing a detective's open mind. They're observant, knowledgeable of trends and correlations but don't let prejudice cloud judgment.

Expensive failures are worth extra effort; failed pressure tubing, turbine blading, dynamic equipment, generator windings, and structural support elements cause major financial outlays to repair. Realistically, repair costs are the least of worries for operating companies when they lose assets and income for weeks or months. Business interruption insurance used to cover these losses, but since the terrorist attacks of September 11, it's become almost prohibitively expensive to maintain.

Explicit basis

The explicit basis formally documents the failure relationships within a planned maintenance strategy. A formal basis helps track failure evolution on major capital equipment like boilers, turbines, or regulated equipment over time. Complex failures that evolve with high cost implication, condenser tube leaks, for example, benefit from a formal basis. As strategies change—like style, design, and composition of installed condenser tube plugs, policy on tubes staking, or other long-term strategic decisions—all parties to complex evolutionary operating-condition management have open access to the plan.

Most failures (including those addressed by codes, licenses, and standards) eventually require technical experts to establish a control strategy that is sustainable over the long haul. In these cases, the explicit basis documents final-code interpretations, technical conclusions, and even engineering analysis, which supports the final implemented PM task selection.

An explicit basis is a godsend to the plant or maintenance manager faced with a regulator questioning maintenance practices following an operating upset. Without this basis, the post-mortem reviewer's presumption is that somebody missed a call.

Take for example a recent case in which a 1300-MWe unit condenser tube leaked. Upon review, it was determined that the apparent PM oversight was a failure to inspect/confirm installation tube integrity of thousands of plugs only one of which failed. The investigation also discovered that the last tube plugs included an elastomer, which introduced a new failure mode! The elastomer rubber rotted and blew out, reopening the original leaks. In hindsight, a better solution than the repetitious all encompassing plug review would have been taking opportunity samples of the new plugs after a few years to validate the elastomer performance. Age-qualifying the elastomer for the high-temperature environment would have been better yet. Time-based replacements could have been used (if needed) based upon projected service life. Or the design may have been rejected altogether. No amount of analysis can compensate for damaging or installing the plugs incorrectly, however. That's maintenance performance.

As with so many plant developments, a solution is selected at the time a problem develops. It solves the immediate problem, but long-term consequences aren't known until many operational years go by.

Every new design or new material carries risk. Experienced engineers know that designs gradually improve over time. They often do so by subtly shifting failure modes. Failure to appreciate a subtle shift in requirements as substitutions are made introduces more potential for surprise. In the case of the condenser tube, the ability to inspect the installed plug elastomer externally was impossible. (Previously, the installed plugs were permanent; integrity was a simple visual inspection ring test.) The new design's saving grace was the ability to replace an installed plug easily, making replacement a snap.

Change basis

Equipment maintenance programs change with experience. Justifications for changes reflect hard-won learning about installations, new materials, processes, and equipment. Changes should capture reasons, developing a running account of program evolution with time. Interval extension uses new materials or experience for advantage. Modified or substituted components can incorporate newer parts, materials, or methods. Different monitoring technology reflects new techniques or even fundamental theory advances. All improve PM methods and support longer intervals and lower cost.

Traditionally, changes were simply made to content, supporting riders, intervals, and methods described in PM work-order workscopes. Since theses were based upon memory, it's easy to see why some changes were made, reversed, remade, and so on. These changes happen as a result of new part availability; if bad experiences occur or vendors go away, equipment necessarily reverts to a previous type or brand. Capturing evolutionary paths is helpful to understand non-linear program changes.

The change basis is a running log of all PM changes and their justification. It's chronological like a diary. Its primary use is to provide reference notes that record why a given change was made. Some regulated programs require explicit bases. Whether explicit bases are required, experience dictates that they are expected and, when available, quickly allay fears about program controls. From an engineering perspective, like a running log—they help maintain and organize notes that support PM program development.

Historically, change bases were difficult to maintain, for engineers kept *yellow dog* notes in their desks supporting plans, intervals, materials, and techniques. Since reliability engineer positions rotate, formalizing notes as retrievable records helps sustain the program.

Component failure modes

The multiple ways components fail can be summarized by component type. Pressure containment service valves can leak; parts subject to abrasion or erosion can wear exceeding dimensional limits. Valves can fail to isolate, or can *weep*. Component set points on safety valves or warning alarms can drift outside limits in mechanical or electronic applications. Multiple parts can lead to the same component failure mode. Seals leak, bodies crack, gaskets harden and leak; many other similar things can cause external leakage. A component failure mode is how a component fails, viewing the component as a "black box." Part failures further define component failure modes into useful, discrete failure mechanisms for craft workers to address.

Parts

Parts constitute components. The natural partition progression from system to train, skid, and component ends at the part level. *Part* is the lowest useful component division. A part can be reworked or repaired as a discrete unit. Parts correspond to component O&M manual takeoff lists. Components incorporate tens to hundreds of critical parts, and critical parts have dominant failure modes. Rarely are more than 50 unique parts *critical*. Most parts are inconsequential; parts with dominant failure modes are the parts of interest for analysis. Identifying and relating parts to failure modes focuses on the parts affecting component functions. Relating failure symptoms to serviceable parts and their failure mechanisms helps perform diagnosis and maintenance.

Failure mechanism

Part-level failure modes are defined as failure mechanisms to distinguish them as sources of component failures and to facilitate their service. Failure mechanisms include mode and cause.

At the part failure level, physical details identify failure symptoms, progression, and effects. A failed part can be identified and uniquely selected. Where critical, hidden-failure symptoms can be sought through instrumentation, predictive monitoring, or calibration. Failure criticality attains meaning at the part-failure level, for parts are the focal points of rework or repair.

Hardware failures affect parts and propagate upward into component, train, and ultimately system function failures. Criticality has little meaning without expressed failure and affected functions. Knowing failure-affected functions—passive and active—establishes criticality.

With a failure mechanism identified, alternative materials, construction, or even capital equipment with longer service life can be considered. Reactor vessel weld cracks, steam generator stress corrosion cracks, turbine blade fatigue cracking—all can be evaluated from engineering life perspective for alternative designs with longer life. Failure mechanism insights help plan future parts replacements, redesign, and corrective maintenance. Turbine rotor-bore inspection, for example, can be selectively applied and tracked across turbine fleets.

Explicit failure mechanism connection allows association of secondary damage and local effects to primary failures. A safety valve failing to lift within limits or failing to reseat following lift has consequences. Secondary effects and progressive failures are more serious part failure risks. Knowing this helps manage overall component system and facility risk. Secondary failures deserve special consideration. Accidents with large losses reveal secondary failures as the *chained causes* that allowed minor events to grow into major ones. Understanding and avoiding secondary effects must rank high on the overall objective list of RCM.

Grouping

Tasks blocking (one component tag)

Work on equipment must be combined at reasonable intervals to minimize trip time. Task blocking that integrates practical work requirements provides a method to work more efficiently around equipment availability.

Blocking assembles tasks to perform at common group intervals while retaining individual task failure-based intervals for reference. Failure-based task intervals are the perfect intervals one would use without blocking. Because re-blocking and replanning work occurs routinely, retaining the original intervals helps re-plan work. It adjusts grouped tasks to operating cycles, seasonal calendars, scheduling intervals, and other natural work cycles that match maintenance performance to equipment availability periods.

Working on specific equipment requires tagout clearance. Equipment tagout is a work overhead, along with trip and tool crib time. Overheads raise final maintenance costs, making grouped work coordination desirable. Common workscope tagouts help to group tasks. Consider a main turbine overhaul. Total scheduled maintenance represents 40 (or so) fundamentally different scheduled maintenance tasks. One could be checking final-stage blade profile erosion, another performing turbine-stage efficiency tests. The former requires shutdown and dismantling under turbine tagout clearance during an overhaul; the latter requires in-service operating conditions (controlled and steady) during prescribed test. The workscope requirements are mutually exclusive; they require separate workscopes to perform at different times under different conditions.

Workscopes accumulate work that can be performed under common conditions, controls, and skills in a convenient administrative group for labor and cost efficiency. A workscope is equivalent to a project management activity. An activity represents one unique group of tasks differentiated from others that requires resources and can be unambiguously identified as *complete* when done.

Across tag workscopes

Rounds and routes perform work across many equipment tags, observing many different equipment pieces over a brief period. Rounds include equipment monitoring by an operator on a short, periodic basis. Daily shift rounds are the prototypical example, but round intervals may be as long as a month. Operators, shift personnel, and even engineers perform rounds. Routes differ slightly involving one skill doing essentially identical work many times over. Vibration data collection on similar equipment, repeated with the same instruments over and over constitutes a typical route. Rounds are usually operations tasks; craft technicians typically perform WO route tasks. Routes are craft task production lines (*see* Figs. 4–13, 4–14, and 4–15).

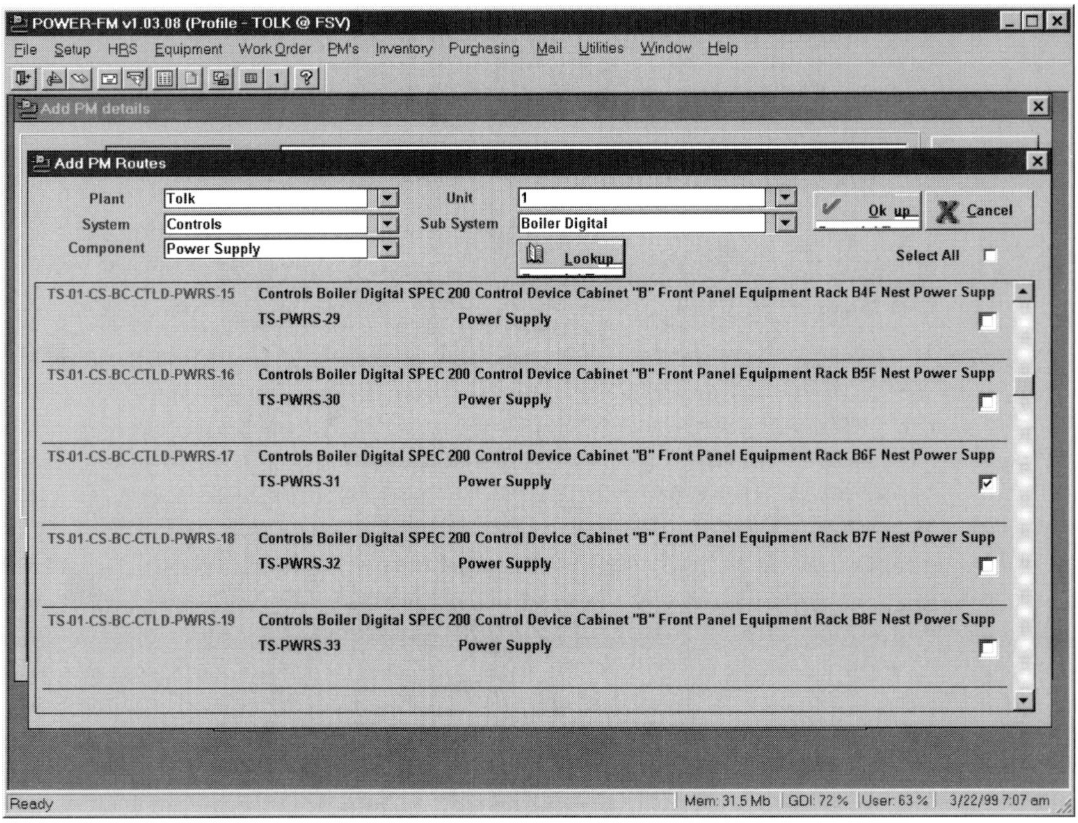

Fig. 4–13 CMMS Workorder Route

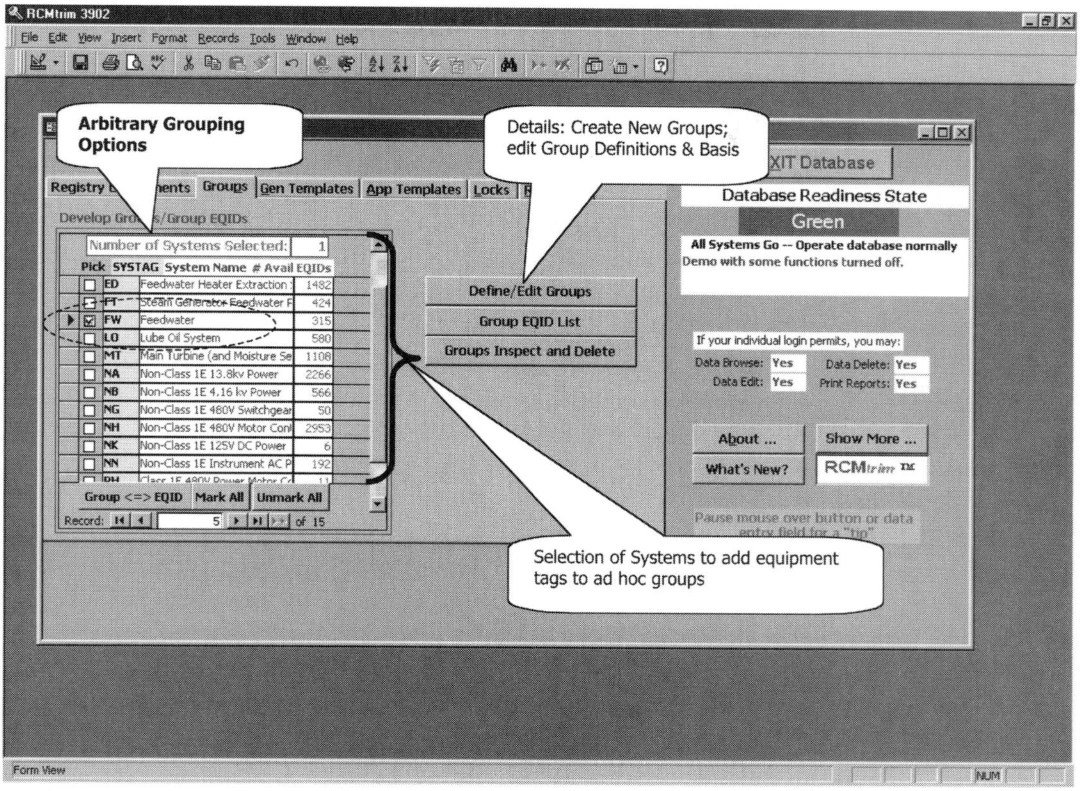

Fig. 4–14 Arbitrary Equipment Grouping

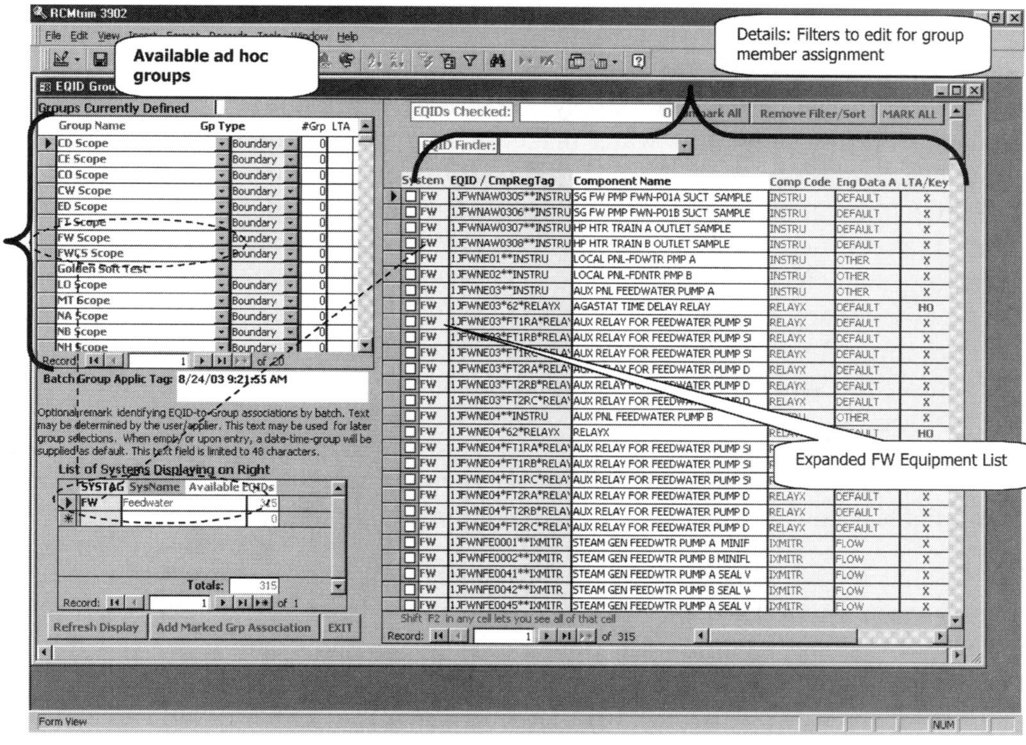

Fig. 4–15 Equipment Assignment/Removal from Ad Hoc Groups

Normal Model

Concept

The normal model envelops equipment with identical plant contexts. Recall that the defining requirement for an applied template is context. Where two or more equipment tags share an identical context—service, risk, and environment—there is no difference in their applied templates. The normal model replicates one primary equipment-applied template to all other members of the same context group. For simplicity and emphasis, the normal model identity comes from the plant equipment tag (and name) of the first member of the normal model group.

Identical plant situations should share one normal model. For example, all level-control loops might be similarly applied on many pieces of common equipment. Where one applied template for a representative group of level controllers models the same standard applied program, that program can be referenced and reused on other identically situated pieces of equipment, even without plant symmetry.

A more common example of a normal model would be multiple identical equipment trains that support a common function. Three condensate pumps that provide boiler feed suction or several transfer coal belts that supply tripper feed or multiple raw water make up pumps provide common examples. Identical trains or capacity (which can be provided for system redundancy purposes) when situated and operated identically, create normal model application candidates.

Conversely, where the situation appears symmetrical and identical but usage is uneven, the normal model does *not* apply. For example, a plant has four river water makeup pumps lifting circulating water makeup 110 feet from a river water source to cooling ponds. Pumps operate in sequential starting service. Operations always runs pump A first, then add B, C, and D in sequential order as cooling tower makeup requirements grow. Because pump A sees the most service, its parts wear out fastest. Programs for pumps A, B, C, and D could hardly be said to be the same because their operating contexts are different. However, after developing one normal model and program for pump A—the wear-out pump—the programs for pumps B, C, and D can be substantially the same. Thus, two normal models address four pumps: one for A, and one for B, C, and D (*see* Fig. 4–16).

Every normal model is associated with one applied template. The normal model provides a plant hardware view for its corresponding applied template.

The normal model concept developed from realizing that working with applied templates invariably modeled one specific plant equipment tag—a *model* hardware piece that had tangible qualities. On early projects, abstract nomenclature identified plant equipment applied templates. Engineers and analysts developing applied templates tried to provide them with a unique identity. Analysts would name and rename applied templates for clarity. After much confusion, realizing that they were simply modeling real equipment, they gave up on arbitrary naming designations and

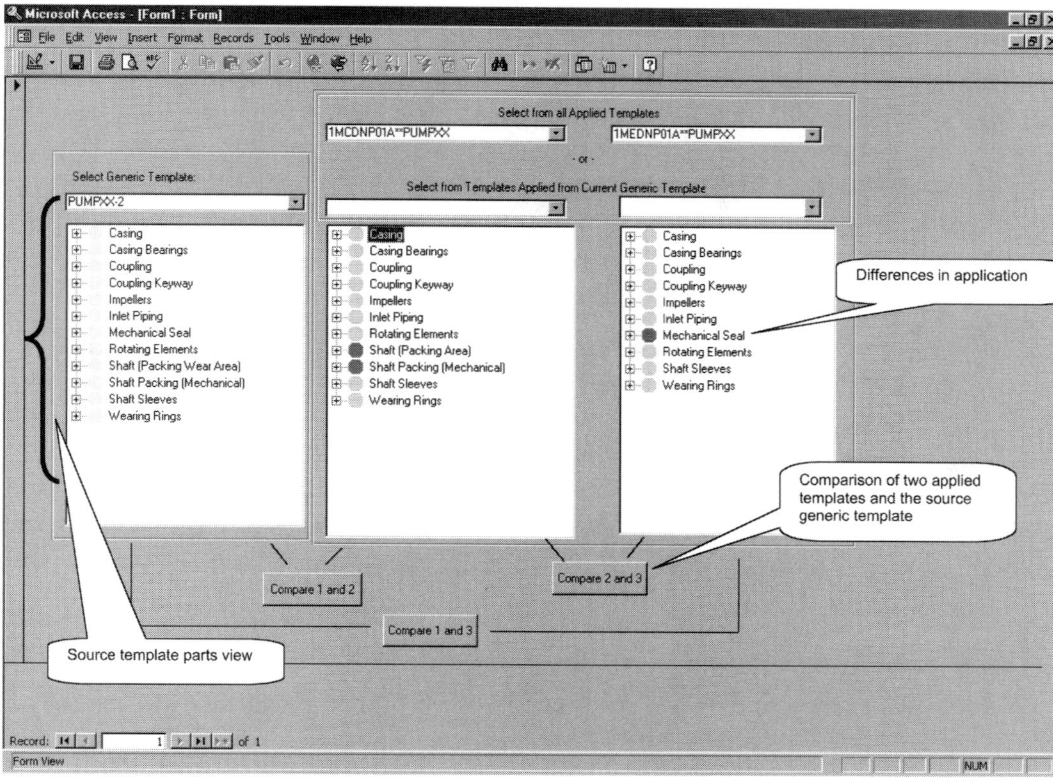

Fig. 4–16 Applied Template Difference Comparison

identified the applied templates simply by the first equipment tag number modeled. So if "Unit 1 Condensate Pump A" was the first component to receive the applied template—1MCDNP01A**PUMPXX—then "Unit 1 Condensate Pump A" was the component tag and name, and the applied template that modeled this pump was simply "1MCDNP01A**PUMPXX." They had created the normal model.

Real plant equipment names and tags made normal models more real and useful, clearer in the users' minds, and unambiguous in any context with respect to equipment modeled and reference source (*see* Fig 4–17).

Applications

The normal model applies when the hardware represented has common context and the same dominant failure modes. This last condition implies that a control-loop normal model can be quite generally applied to many control loops, provided the failures, risks, and context are the same. Predominant control-loop failures are *drifts* and *fails to operate*; the exact loop nature is immaterial as long as the risk exposure is the same. *Drift* is the gradual change in output from setpoint in electronic circuits with time due to variations like conductivity changes, temperature changes, etc. *Fails to open* is the sudden failure ("open circuit") from a discrete part failure, loose lead, or other

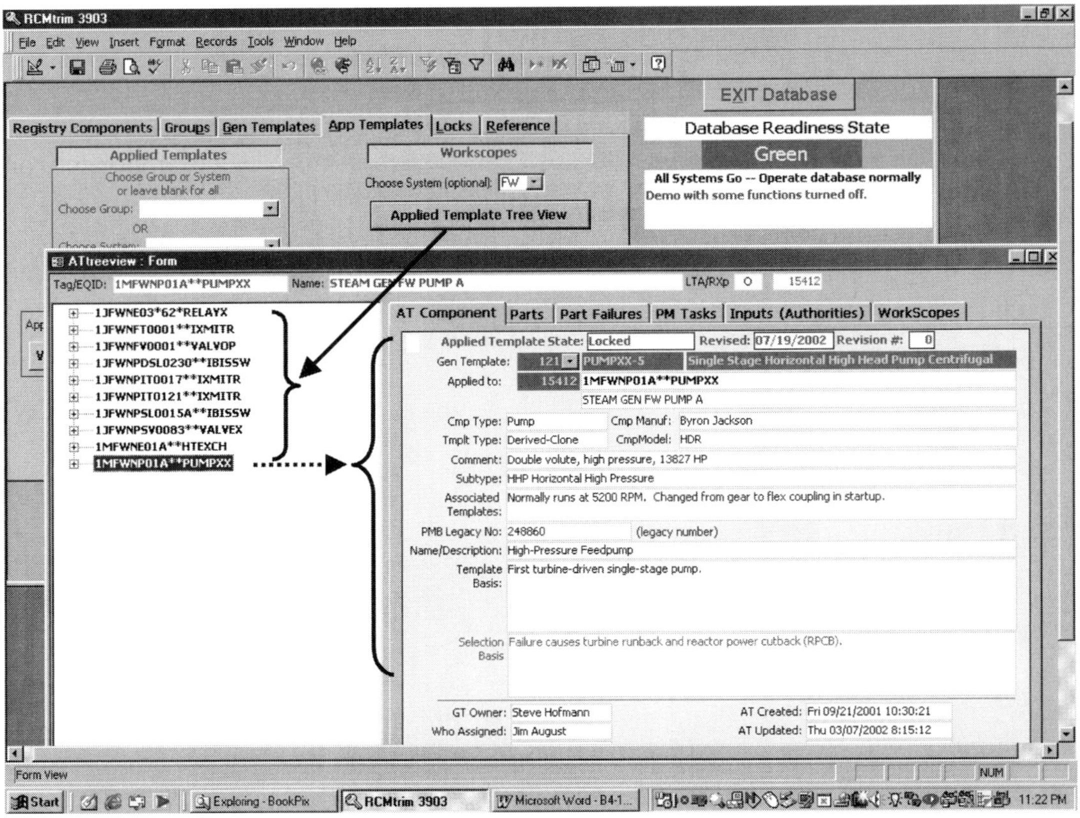

Fig. 4–17 Normal Models: Feedwater System

continuity interruption. The normal model validates standard calibration instrumentation and control programs. It justifies this standard Instrumentation & Control (I&C) template and explains why calibration is valid across a broad instrument control loop application spectrum—identifying criteria that establish model limits. (For I&C components, calibrate and channel check address most failures. This is also the standard PMO-applied template, only at one interval.)

Instrument loop

The instrument loop provides a special template case. Many electronic components exhibit two predominant failure modes: *failure to operate* (open circuit) and *drift.*

Special cases include time-based failure mechanisms of any subcomponent, especially sensor elements and elastomeric seals. A sensor immersed in a fluid stream (like a pH sensor in a condensate stream) requires periodic cleaning. The cleaning activity can be addressed during the periodic calibration within the loop or as special-failure mode under loop drift. The loop integrates individual component and parts failure modes under one task—a significant benefit for scheduled maintenance.

Many failure modes can be integrated and dealt with as a common scope, improving PM efficiency. Of course, when failure occurs, experts must analyze the composite elements to locate the culprit part, but the grouping strategy takes advantage of the relative infrequent and random complex equipment failure to check functions and only perform work as needed. This strategy works well for active components or components that can periodically be activated to test and calibrate. The channel check/calibration combination reflects this concept. If a channel can't be checked successfully for continuity, it can't be calibrated.

Electronic instruments and their housings have many passive functions such as sealing out the environment. Power generation environments include moisture, vibration, signal noise, voltage spikes, thermal extremes, and heat-up/cool-down cycling. These variable random stresses contribute to random electronic failures. Protection from environmental stress provided by housings, covers, thermal inertia devices, and other design features can be compromised by environmental changes. Where standby instrumentation must function under adverse accident conditions, PF-interval, as well as latent aging effects of accident and other stresses must be accommodated. This rationale is the basis for 10CFR50.49 the rule for nuclear "environmental qualification" (EQ) programs.

When an age-based failure occurs in electronics, aging must be factored into the overall loop template. A component in a loop for control, alarm, or instrumentation purposes may be easier to treat out of the loop for aging, seals, gaskets, or even electronic components if these effects aren't evident in test or calibration results. Power supplies and electrolytic capacitors provide two cases in point. Without closer inspection, the loops they support can be active and functional even though the underlying electronics parts are aged. Only diagnostic evaluation shows the ripple output of a capacitor or power supply. Aging stresses could lead to failure in the design basis event during which the electronics instrumentation and control signals must perform.

Trains

Trains are groups of equipment that replicate common functions. Trains reduce to a set of normal models, replicated in each train. For the purposes of this discussion, trains are symmetrical elements. They speed analysis by allowing the analyst to develop a solution for one train and replicate that solution for other equivalent trains.

Many plants use trains in their design. Trains save time during design and operation, a veritable force measure. Their analytical solution can be replicated many times over for capacity, redundancy, or both; the basic engineering remains the same.

Cases where asymmetry in nearly identical trains occurs have special engineering interest. Asymmetry provides special customized functions. For example, one loop of a condensate pump train may supply makeup for the control rod hydraulic drive system; condensate, which ordinarily plays a relatively modest safety risk role, does double-duty in this case with a second, important safety role. Asymmetry is introduced.

The train that supplies control rod makeup now has more safety significance. Trains with asymmetry of any sort deserve special consideration to see whether asymmetry raises or lowers their risk exposure rank based upon the special role.

Skid

Skids are the mechanical analogues to control loops. Mechanical subassemblies are built up as skids. Most skid equipment supports a subsystem. Where the equipment can be grouped and tested together—as for a fair amount of standby equipment—simplification is achieved by treating it as a skid.

Like the loop, when a subassembly or component exhibits aging that predominates separate from the skid, the aging can be explicitly addressed as PM associated with that component tag (*see* Fig. 4–18).

Sub-partition

Identification schemes code equipment tags so that similar elements in each of several different units differ only by a unit prefix, or they can be nearly random. For example, 1MCDNP01A**781230 and 2MCDNP01B**781230 are the nearly identical A and B condensate pumps for Unit 1 and 2 condensate systems, respectively. Some coding schemes stop at a high level, while others add great detail. Some coders followed PID drawings to take off components along process paths using systems, trains, skids, and other very systematic methods. Others worked with inconsistent rule sets, which plant tags reflect. Equipment coding reflects administrative guidance of the AE's engineering department. AEs have standard consistent coding schemes that are like a signature; from the way a plant's equipment is coded, the engineer can surmise who their AE was.

RCM analysis must use prior work, and an ideal RCM process must deal with either extreme—too much, or too little detail—as well as everything in between. Too much detail can be corrected by using primary equipment associations to group equipment with no scheduled maintenance needs. Adding new equipment tags in the plant database or partitioning the plant's equipment tags for RCM corrects too little detail.

Sub-partitioning equipment with significant subcomponents extends the component tag list analysis to the next lower level, which often improves template-based RCM analysis. The appropriate coding level is never known exactly until analysis is underway. It is desirable to have an equipment tag target for every component that could become critical in the equipment partition and receive a normal model-applied template. Coding and partitioning should identify the partitioned subcomponent type so a generic template type can be easily, uniquely associated. Turbine skids, sub-partitioned to identify *lift pumps* on the skid with no tag, ideally identify a lift pump as a *vertical centrifugal pump* (e.g., pump, vertical centrifugal). Partitioning components into subcomponents with dominant failure modes should not alter component tags that might receive an applied template.

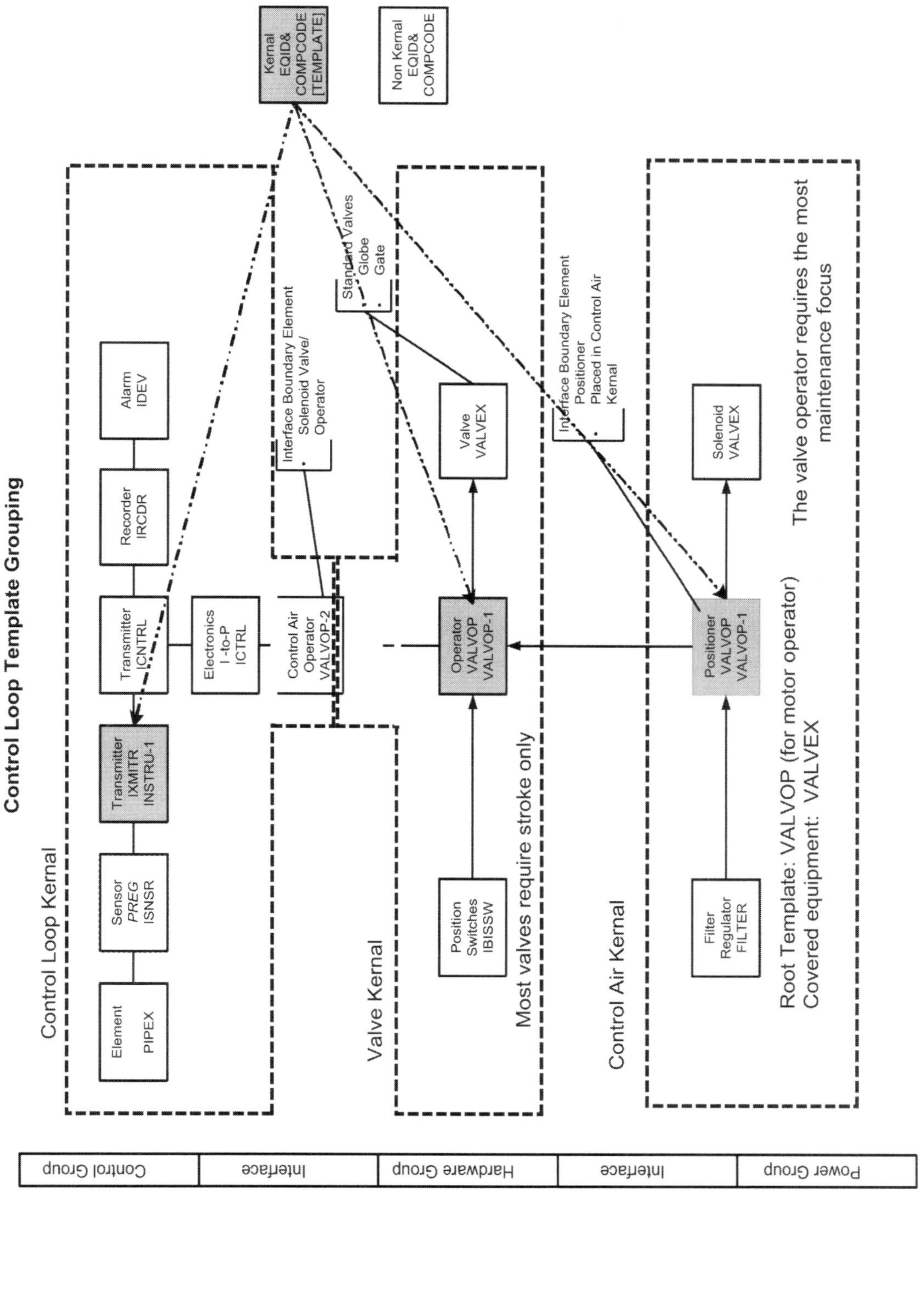

Fig. 4–18 Control Loop Grouping

Problems

Normal models can be used too broadly—where they don't apply. I&C loop-calibration normal models are sometimes overextended. It's too easy to presume that some unknown loop behaves like the standard loop model, without doing the validating research.

Consider an electronic loop with an aging part. Given similar standard control loop template application, when the aging part life replacement task interval differs from that for loop drift aging, component service periods are either missed or must be separated for incorporation into two workscopes. *Calibrate loop* could address drift, and *replace seal* addresses the aging soft parts for environmental enclosures. For pH sensors or Na analyzer loops with a calibration interval applied, if the sensor ages with a six-month cleaning life and the normal model loop calibration and channel check is two years, either the sensor doesn't get cleaned, or an applied template reflecting the sensor cleaning WO workscope is required. *Canned* I&C templates can't manage component aging based on their simple modeling. However, I&C normal models are easily extended to cover this case. They simply require a new template, with a new workscope (*see* Fig. 4–19).

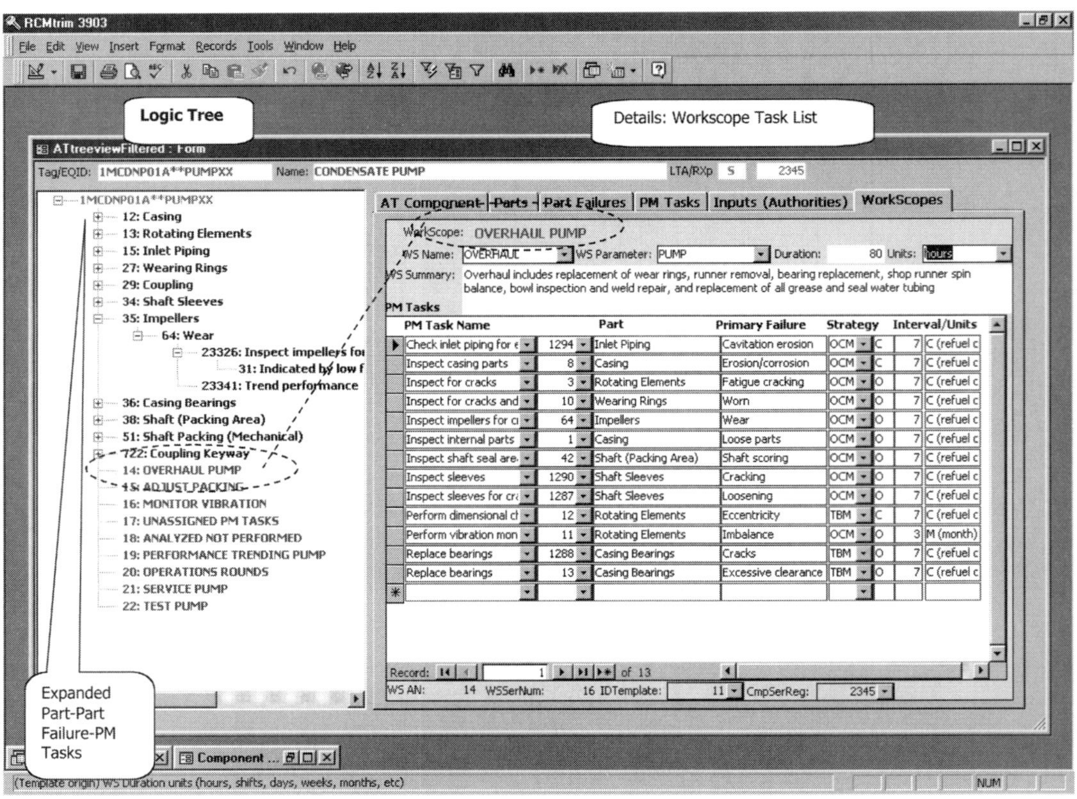

Fig. 4–19 The Product of an Applied Template: A Workscope

System Templates

Concept utility

Systems can also be modeled for generic application as a sort of global template. Nominally, various plant systems occur identically in many applications: Virtually all Rankine cycle plants have a condenser, condensate pumps, and condensate monitoring instrumentation that are similar in design from plant to plant. The same can be said for combustion turbines and combined-cycle units. Condensate systems vary among them in style and design of plant—large BWR/PWR condensers usually exhibit features different from fossil supercritical boilers—but the general designs are still similar. Using a system template speeds the development of a BWR condensate system (as well as condenser) by modeling the design upon another one and making appropriate adjustments (*see* Fig. 4–20).

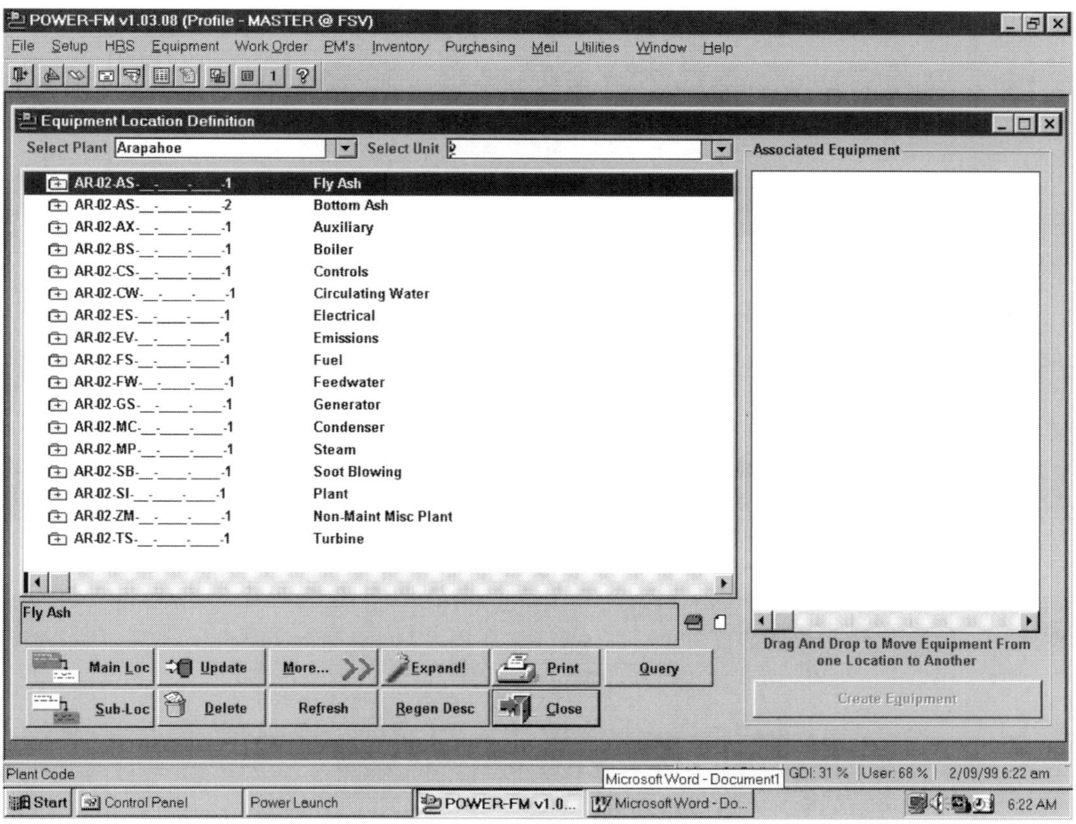

Fig. 4–20 CMMS System Templates

Requirements

Creating system templates requires development of the nominal flow processes, critical equipment, useful generic models for the critical equipment, and the normal model. The risk exposure map for the representative system with basic flow processes, skids, trains, and risk classification (SOC) with basis provides the rough system-generic template. To be useful, the generic components in the system template need to be modeled specifically for that system. This requires an applied template for each normal model in the nominal system template. System template-supporting generic templates can also be used (*see* Fig. 4–21). Once the basic system process template has been developed, the primary adjustments from the reference system model involve similar trains, equipment, and the level of redundancy they provide.

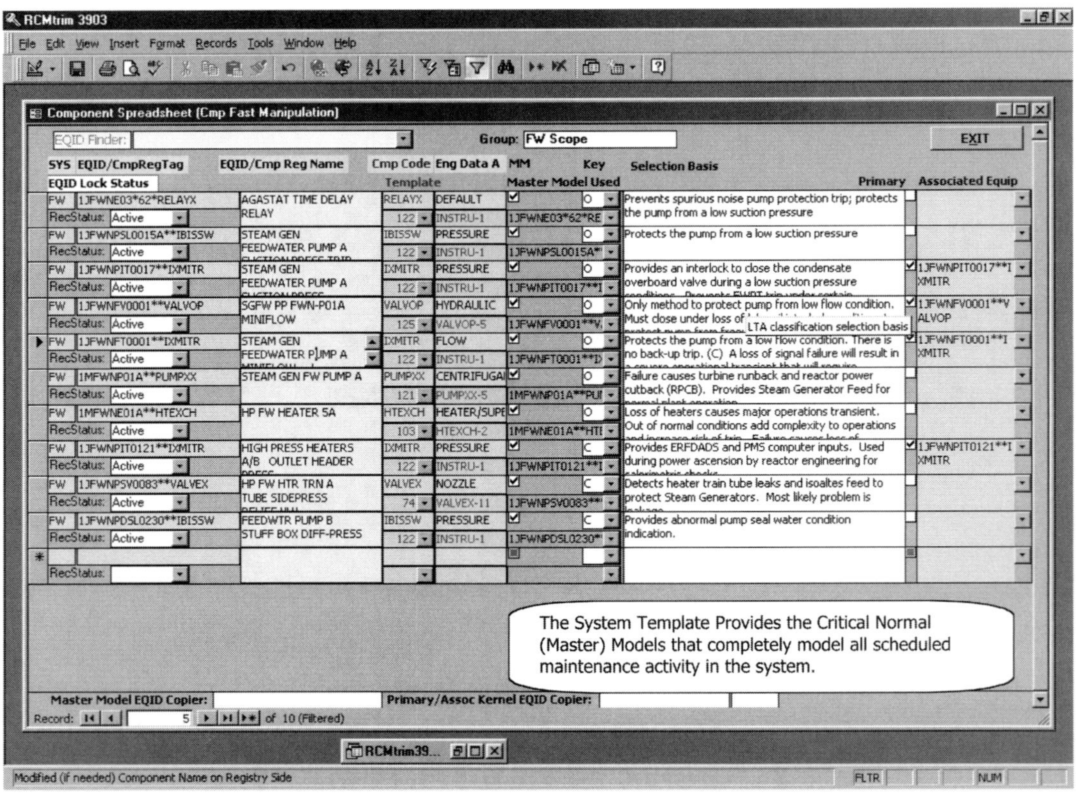

Fig. 4–21 System Template: Feedwater

5 Component Failure

Context

Failures occur in many ways and in many contexts. Failures start as bottom-initiating events, some of which propagate upward causing system failures. Reliability maintenance strategies do not limit failures outright, for some are inevitable, based upon randomly failing subcomponents in designs. Rather, the goal is to *manage* failures within limitations and intent of design (*see* Fig, 5–1).

For most maintenance participants, acknowledging what they all privately appreciate—that failures are not all created equal, or necessarily all bad—comes hard. As operators know, some failures are worse than others.

Risk analysis is RCM's first feature, one to which maintenance people have less access compared with operating staffs. At a high level, operators have a working understanding of failure based on risk while maintenance workers understand and interpret equipment failure at the equipment level. Even here, mechanics (unlike operators) don't keenly grasp equipment operating risk. Risk depends on design installation redundancy, probability of failure, and consequence that compromises function of equipment in the operating interest period.

Equipment carries performance expectation. Falling short of expected performance constitutes failure. Understanding failures in an engineering world requires specifying performance expectations appropriately with performance limits. Performance limits can be expressed formally at the systems level with quantified outputs. Industrial facilities produce tangible goods and services. The inability to produce at specified quantities and quality levels defines failure (*see* Fig. 5–2)

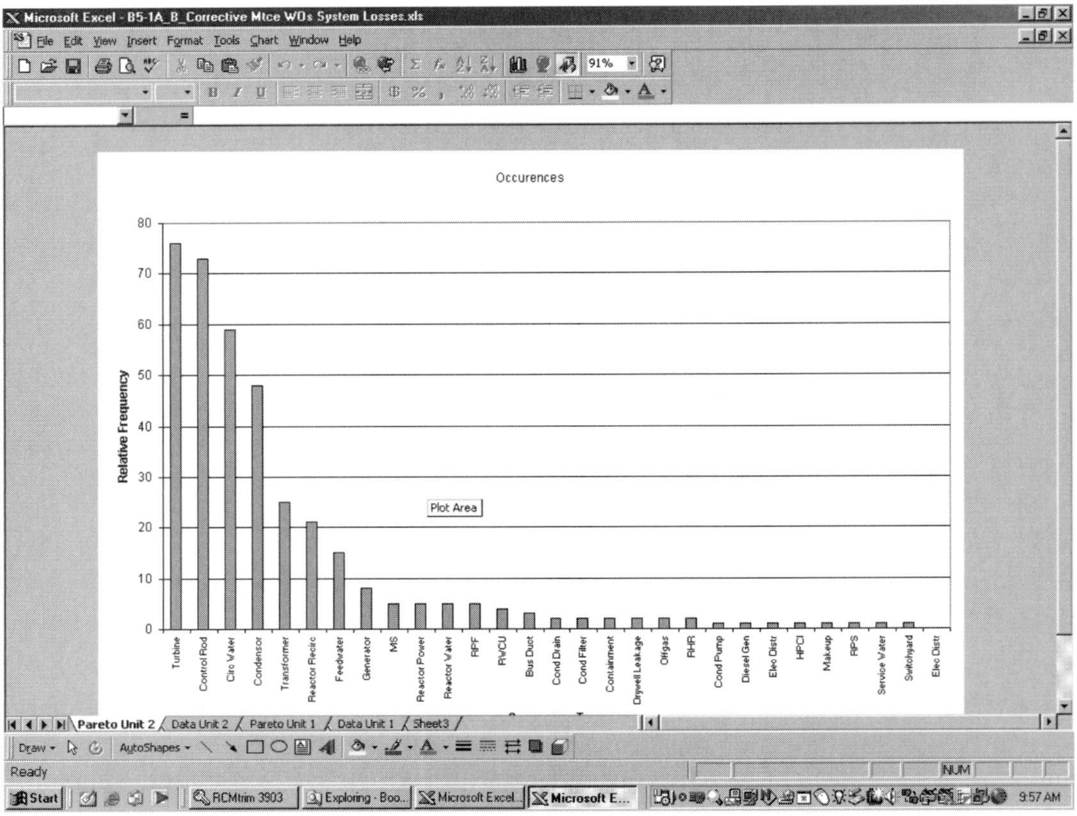

Fig. 5–1 System Losses from Corrective Maintenance Work Orders

Systems incorporate adequate equipment to produce at specified functional levels. *Functions* are system outputs that require valid input as well as equipment internal functional processing. Trains and components also have functional expectations.

Defining failure at *any* level means specifying supporting functions and determining how to ensure the functions are available. Engineers define equipment adequacy with design specifications. Testing by start-up crews assures that it's provided in new construction plants.

Practical failures occur when stress exceeds resistance. Engineering failures occur when a measure exceeds specified limits. Classic metal yielding occurs when loads exceed the metal capacity to carry load. Engineering is a science of designing resistance into parts by material selection, composition specification, environmental controls, and process limitation. Nonetheless, conditions arise when stresses exceed capacity, and failure occurs. Anticipating and designing for stresses with margin defines engineering.

Where failures occur, assumptions and conditions that lead to functional failure must be examined to determine whether stresses exceeded design or design failed to perform to expectations (*see* Fig. 5–3).

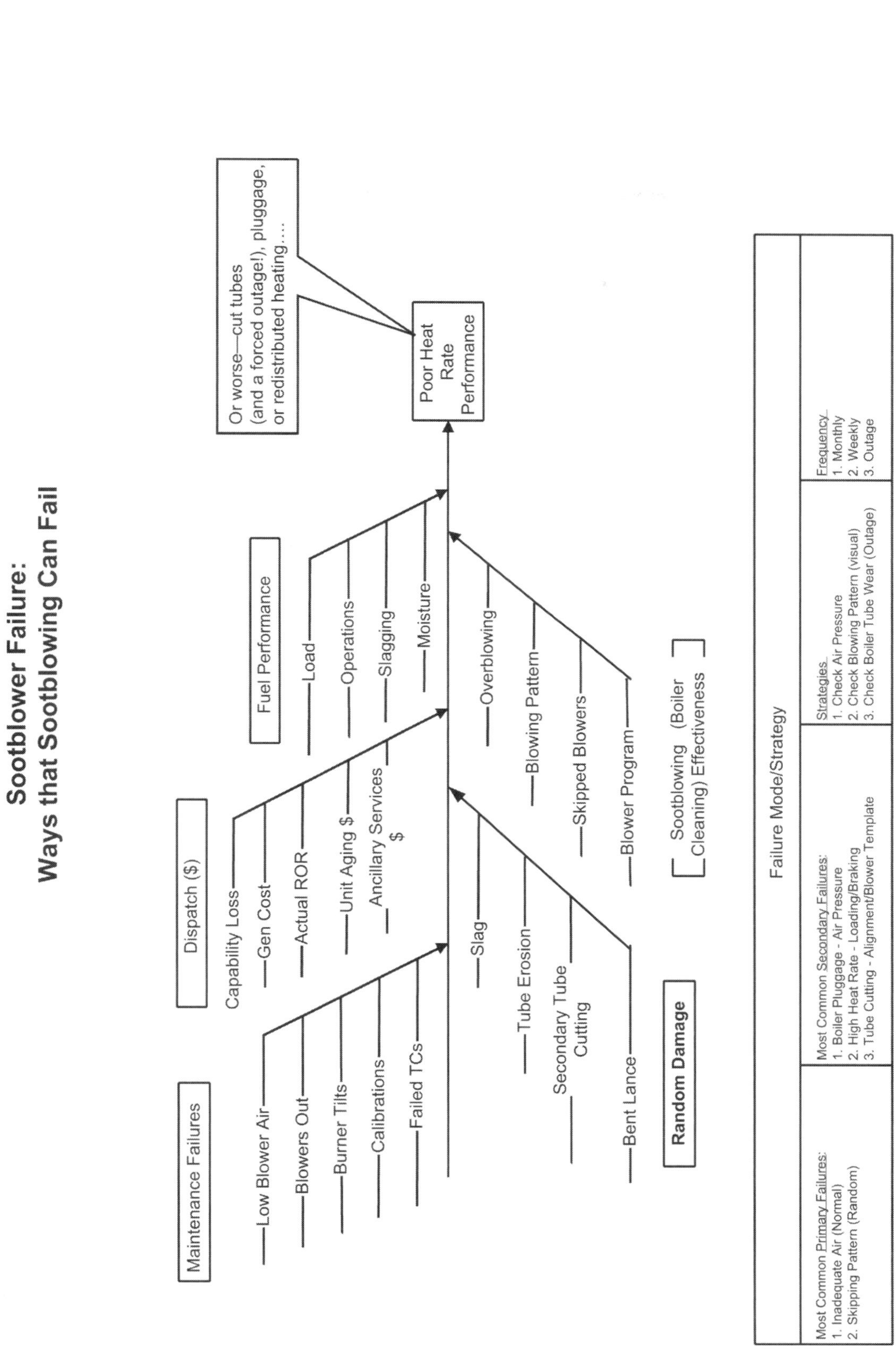

Fig. 5–2 Sootblower Failure: Ways that Blowing Can Fail

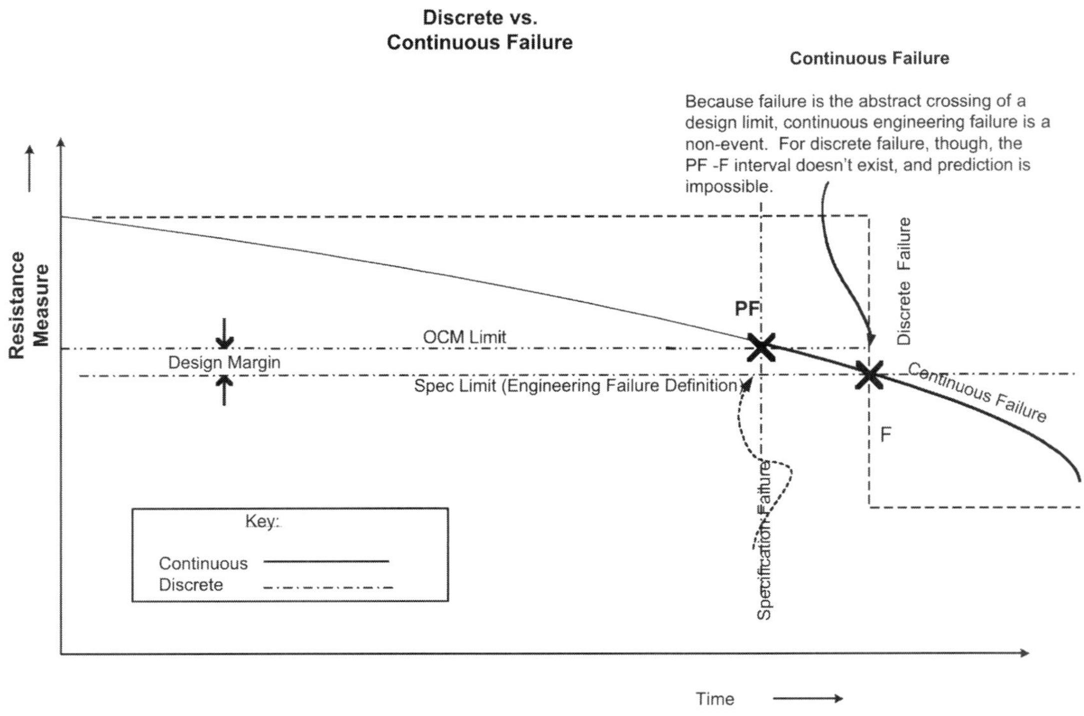

Fig. 5–3 Stress Limit Curve

Failures start with conditions and events. Some are continuous processes, while others are discrete. Engineering failure begins when a variable attribute exceeds a specification limit. The variable is often continuous, but it need not be. For boiler overpressure safety, process pressure continuity is obvious though not explicit: Gradual oil dilution causes viscosity to exceed limits. Consequences will not be pronounced, but it is no less an engineering failure. Recognizing that many failures occur continuously helps understand common variable failure. Even without bells or alarms, failed states become unmet expectations in the longevity and serviceability of equipment.

Failure ambiguity centers on detail and specificity. Defining failed function requirements is an art. Describing failures at the system, train, and component level requires skill and standards. At a high level, industrial systems typically have fewer than 10 functions, and rarely exceed 20 functional failures. Certain functions are present in virtually all systems, like *provide status* or *maintain structure*. Others are very specific, like *convert mechanical torque into electric power (current and voltage)*.

At the component template level, the need to specify failure numerically ends, and operator expectations define failures. Function statements such as fails to start, fails to load, and fails to lift describe components failing to perform as expected. Practically, once failure has been defined, scheduled maintenance must evaluate failures against functional requirements. Even when functional expectations and failure to meet *specs* are evident, failed equipment correction generally involves diagnosis.

Diagnosis reaches inside the functional equipment black box abstraction to understand causes of failure. Restoring failed equipment requires understanding how the failure occurred, which, in turn, involves causes. Although functions define failure, understanding how functions arise allows diagnosis of and correcting the failure. For scheduled maintenance, the focus is on failure diagnosis. Corrective maintenance effort directs fixing diagnosed failures *(see* Fig. 5–4).

Component modeling

Components conveniently fill two hardware classification levels. The upper tier is broadly classified and includes things like pumps, valves, and motors. The lower tier develops subtype-specific applications. Valves can be sub-classified as ball, check, globe, or gate; motors can be horizontal form-wound, vertical synchronous, or induction; breakers can be 4 kV air blast, vacuum- or oil-filled. Two-type levels adequately provide enough detail to locate and correlate generic templates with plant components. Many industrial hardware classification schemes use two tiers. Beyond two levels, the complexity can outweigh additional value.

Basic Failure Concepts

Complexity

Industrial equipment is complex. Complexity provides both advantages and liabilities. Complexity incorporates multiple redundancies, capabilities, and features that allow equipment to provide more functionality and service while providing better status information and less risk. As complexity increases, diagnostic and repair skill requirements also increase. Some equipment installations incorporate more diagnostic components to identify failure and diagnose conditions. Complexity in equipment design hasn't reduced the need for expert diagnostic services.

Designers use complexity to avoid operational problems. Redundancy helps to manage internal part failures, and many equipment controls include redundant design features for control, safety, and alarm components. Reliability engineers must understand complexity, using specific strategies to manage complex equipment.

Complex equipment can be identified empirically from the behavior and nature of the failures experienced. Consider briefly the common PC. The failure that most people experience most often is periodic lockup. Although some high-risk activities and tasks introduce more lockup risk (loading new software, browsing the internet extensively), simple lockup tends to occur randomly. The random pattern of this problem probably best describes the casual user's primary operating frustration. Lockup is a random failure in a practical component application, classic in form and known-well to every user! On first impression, it seems to occur unpredictably whether the user has just started work or has been at it a full day. Consequences of a lockup include loss of work since the last save operation was done.

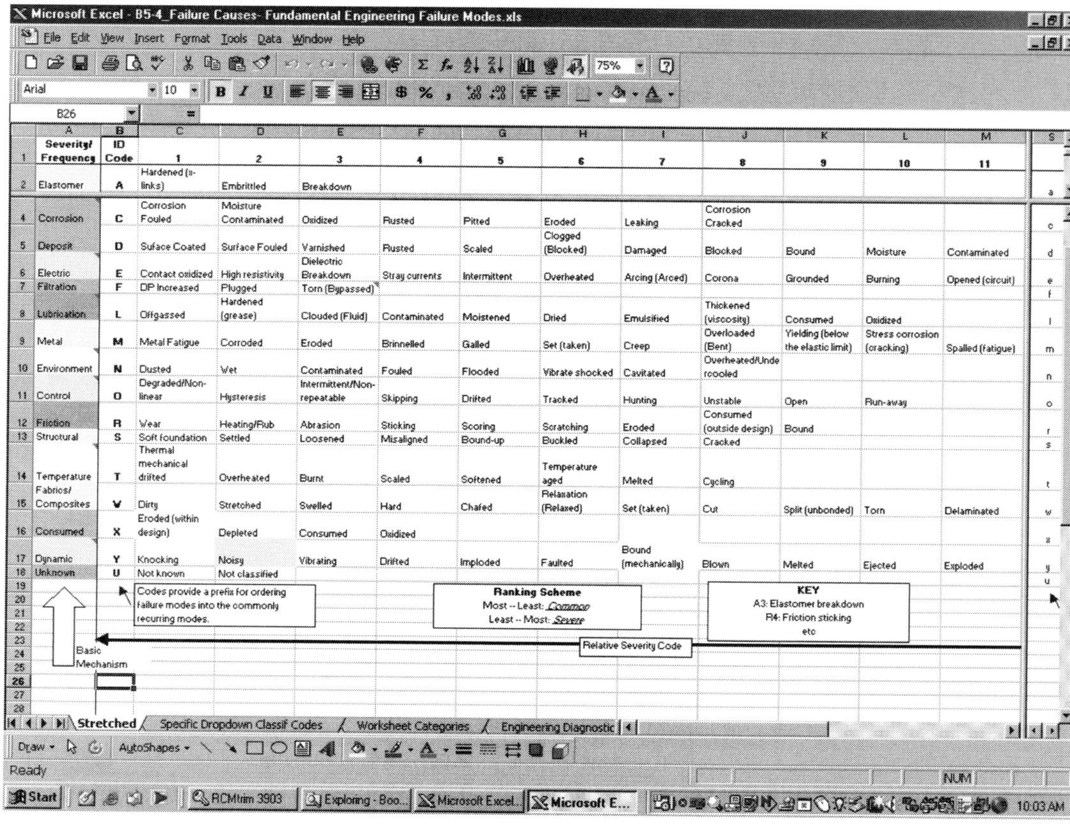

Severity/ Frequency	ID Code	1	2	3	4	5	6	7	8	9	10	11
Elastomer	A	Hardened (x-links)	Embrittled	Breakdown								
Corrosion	C	Corrosion Fouled	Moisture Contaminated	Oxidized	Rusted	Pitted	Eroded	Leaking	Corrosion Cracked			
Deposit	D	Suface Coated	Surface Fouled	Varnished	Rusted	Scaled	Clogged (Blocked)	Damaged	Blocked	Bound	Moisture	Contaminated
Electric	E	Contact oxidized	High resistivity	Dielectric Breakdown	Stray currents	Intermittent	Overheated	Arcing (Arced)	Corona	Grounded	Burning	Opened (circuit)
Filtration	F	DP Increased	Plugged	Torn (Bypassed)								
Lubrication	L	Offgassed	Hardened (grease)	Clouded (Fluid)	Contaminated	Moistened	Dried	Emulsified	Thickened (viscosity)	Consumed	Oxidized	
Metal	M	Metal Fatigue	Corroded	Eroded	Brinnelled	Galled	Set (taken)	Creep	Overloaded (Bent)	Yielding (below the elastic limit)	Stress corrosion (cracking)	Spalled (fatigue)
Environment	N	Dusted	Wet	Contaminated	Fouled	Flooded	Vibrate shocked	Cavitated	Overheated/Undercooled			
Control	O	Degraded/Non-linear	Hysteresis	Intermittent/Non-repeatable	Skipping	Drifted	Tracked	Hunting	Unstable	Open	Run-away	
Friction	R	Wear	Heating/Rub	Abrasion	Sticking	Scoring	Scratching	Eroded	Consumed (outside design)	Bound		
Structural	S	Soft foundation	Settled	Loosened	Misaligned	Bound-up	Buckled	Collapsed	Cracked			
Temperature	T	Thermal mechanical drifted	Overheated	Burnt	Scaled	Softened	Temperature aged	Melted	Cycling			
Fabrics/ Composites	V	Dirty	Stretched	Swelled	Hard	Chafed	Relaxation (Relaxed)	Set (taken)	Cut	Split (unbonded)	Torn	Delaminated
Consumed	X	Eroded (within design)	Depleted	Consumed	Oxidized							
Dynamic	Y	Knocking	Noisy	Vibrating	Drifted	Imploded	Faulted	Bound (mechanically)	Blown	Melted	Ejected	Exploded
Unknown	U	Not known	Not classified									

Fig. 5–4 Failure Causes: Fundamental Engineering Failure Modes

The strategy for guarding against the consequences of a lockup is the same as any random failure—redundancy. Save work regularly, and on an interval that reflects usage, cost, and risk. Maintain backups for more severe, less probable, but still random failures.

Complex equipment model candidates occur frequently in modern industrial applications. Most people are comfortable treating microprocessor electronics as complex because they know little about how microprocessor electronics work. They are familiar with vexing random failures in inexpensive consumer electronic products based on the many microprocessors embedded there. Many mechanical devices exhibit essentially random failures from many different failure modes, thus meeting *complex* definition criteria. The surprising observation is that complex equipment can be empirically defined.

Any equipment that fails randomly with no predominant failure pattern exhibits complex failure behavior and can be treated as complex equipment from a reliability perspective (*see* Fig. 5–5).

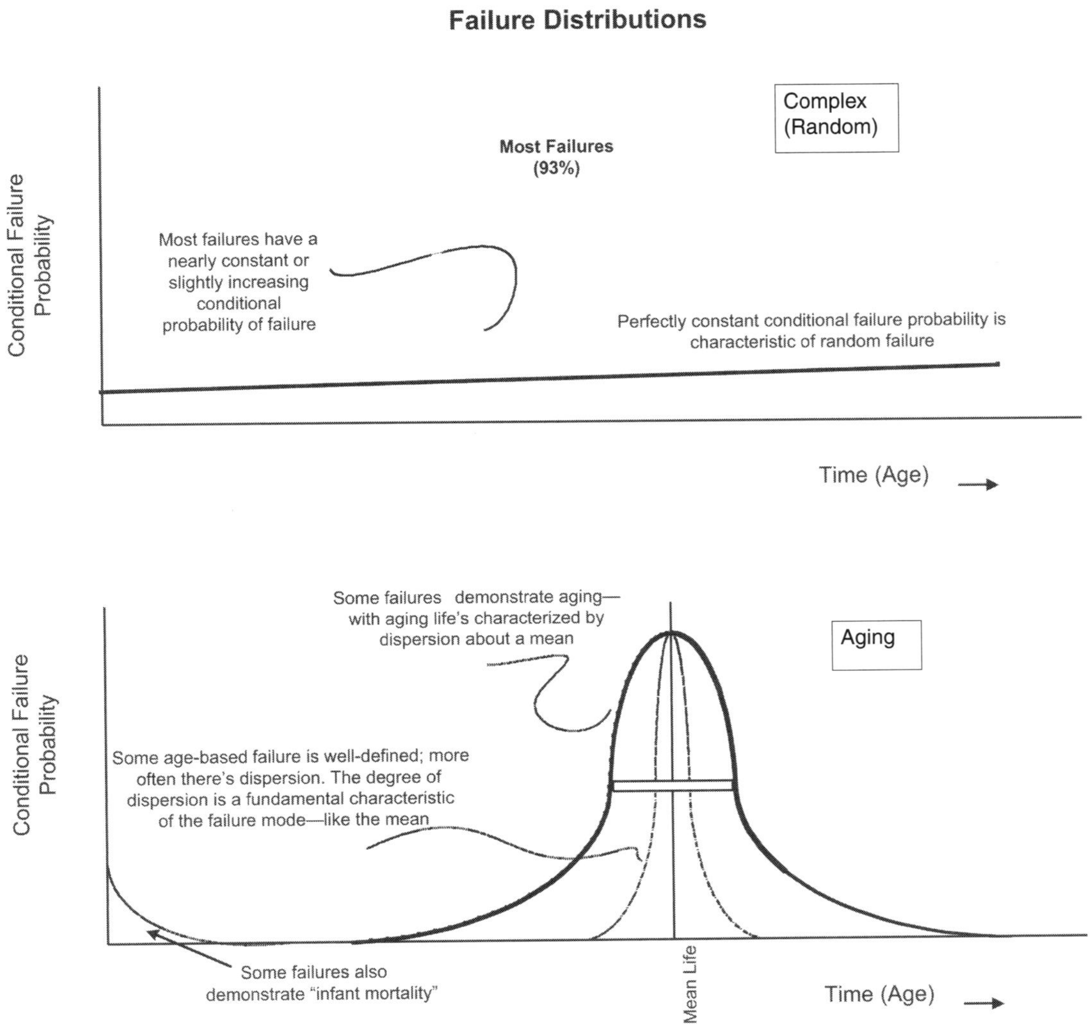

Fig. 5–5 Failure Distribution for Task Selection

Dominant failure modes and *fishbones*

Dominant failures are manageable. Extracting dominant failures from the sea of potential candidates requires engineering sleuthing. Some intelligent guesswork on the engineering failure causes at work helps focus and collect statistics to validate intuition.

From Pareto distributions of failed equipment, Ishikawa *fishbone* diagrams are useful to strategize, organize, and establish how failures contribute to functional failure, as well as establishing those that are dominant failures. (Dr. Kaoru Ishikawa is the originator of the diagram. The name *fishbone* refers to its general appearance.) Fishbone diagrams are helpful in discussing evident failures because they visually provide a structure on which to organize cause-and-effect relationships, discuss patterns, and identify likely leads to validate suspicions for frequent failures and their possible causes (*see* Fig. 5–6).

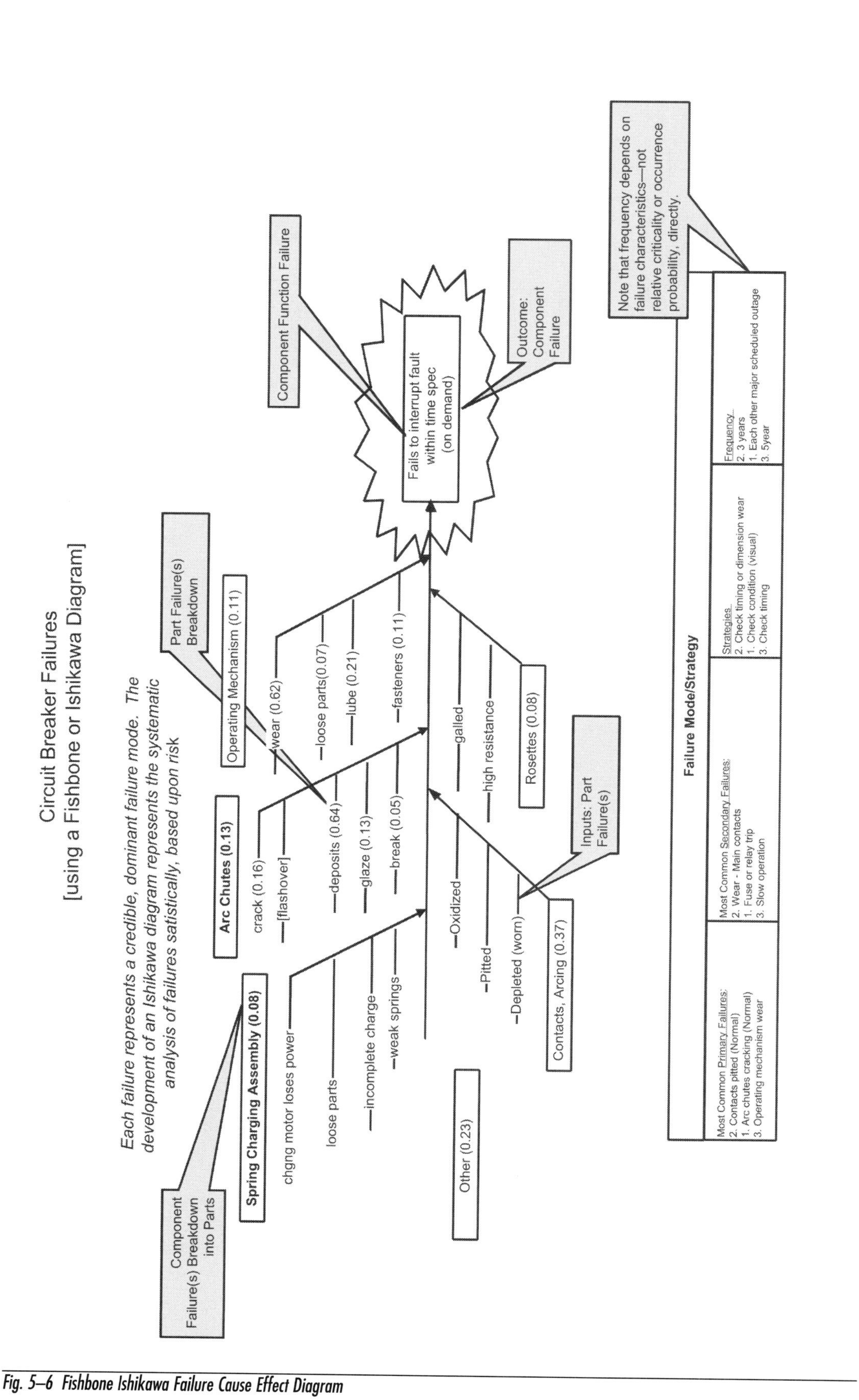

Fig. 5–6 Fishbone Ishikawa Failure Cause Effect Diagram

Failure modes and effects analysis (FMEA)

Failure modes and effects analysis studies failure at the hardware level. While engineers can partition system, equipment, or components to any level—parts fail. Parts are the lowest discrete replaceable unit in a component and are the level at which diagnosis, replacement, or rework occurs.

FMEA examines part physical hardware failure from an engineering perspective. (*see* Fig. 5–7) FMEA is one of the earliest qualitative techniques to analyze component failure, and its application identifies the initiating events that must be considered to understand the origin and consequences of failure.

Consider the following vertical pump failure modes:

- worn impeller blading
- cavitation eroded impeller
- worn wear rings
- blocked suction bell screen
- eroded housing
- failed stage connecting bolts
- eroded cutlass bearings

Suction bell blockage causes immediate pump capability loss. Design strives to manage what is likely caused by an external, random event (e.g., debris induction that could bind and seize the pump rotating parts and shaft). Pump design could include a shear coupling that failed preferentially rather than causing a rotor shaft cracking upon overload. The failure occurs in an accessible location, so rework can proceed with less extensive parts replacement or repair. (Design discussed implicitly presumes redundant capacity is available or the application is not critical.)

Normal wear aging from worn rings and impellers will cause a gradual decline in capacity. Loss of cutlass bearings is likely to increase pump noise to the point of operator pump shutdown before an actual bind occurs. Binding, though improbable, would result in the support drive shaft cracked or bolts sheared from overload.

Shifting focus to the motor, which supports a vertical pump, problems might include:

- thrust bearing loss
- axial bearing loss
- lubricant loss (motor lubricating oil)
- lubricant failure (due to viscosity, contaminant, or moisture limits)

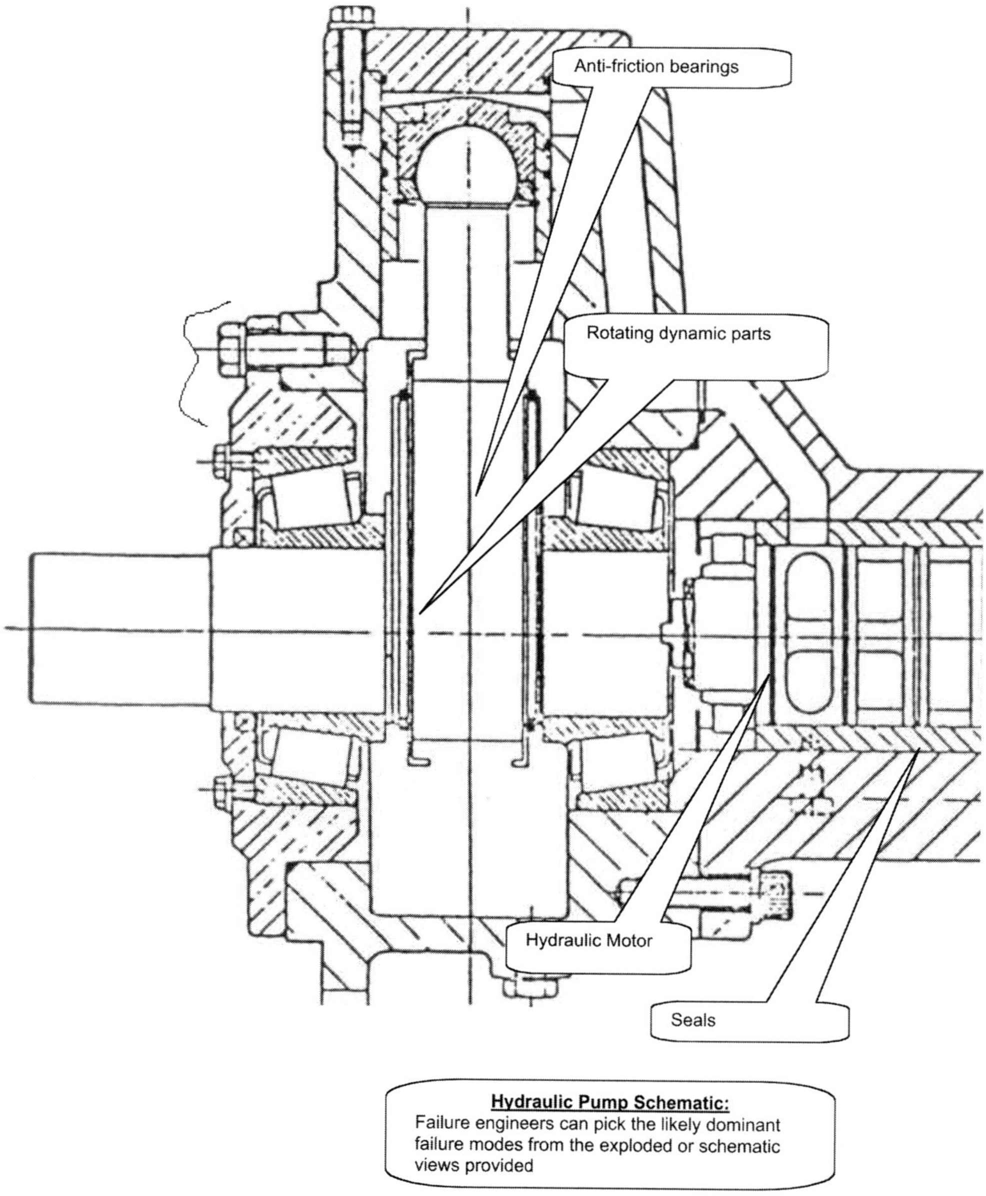

Fig. 5–7 Pump Schematic

The development of practical failure prevention strategy begins with the identification of failure modes at the part level. Some of these part failure mechanisms will be predominant. Many won't ever develop in complex equipment due to intermediate specifications and part controls such as lubrication, which ensure the life and performance of basic parts.

FMEA development begins by reviewing the component supplier's drawings and part lists. Most parts on complex equipment never fail; those that do demand attention. Wearing and eroding parts are easiest to recognize because of fluid exposure (to steam, water, or oil flows, for example). Seals and gaskets, o-rings, and other elastomers also get attention because of their time-based aging behavior. Thermally cycled parts, like fasteners, have redundancy in load capacity, yet failures will occur after many stress cycles. Age exploration should study equipment like turbines that have very long service lives. As the FMEA develops, more and more parts with age-based failure potential appear on the partition list. What remains are progressively larger, beefier components such as housing bells and base plates that rarely fail in common service applications.

These parts need not be addressed on the parts list, even thought they are physically large and functionally important. They don't cause dominant failures. Caution is in order, though. Housings in acidic-water areas (like mine reclamation geography) exhibit high general corrosion rates, and these housings *do* fail. Base-plate anchors for grouting can fail in locations subjected to high shock vibration levels such as around heavy ball or rolling mills. Oscillation pounds concrete floors into powder, weakening grout and anchor hardware over time.

Hypothetical FMEA development is an academic exercise. Statistically, many failures on one piece of equipment will never be seen in one specific industrial environment. Efficient engineering requires identification of failures that will emerge for appropriate attention, while ignoring those that won't. All the while, caution and objectivity are in order where there is an impact on safety or operations (*see* Fig. 5–8).

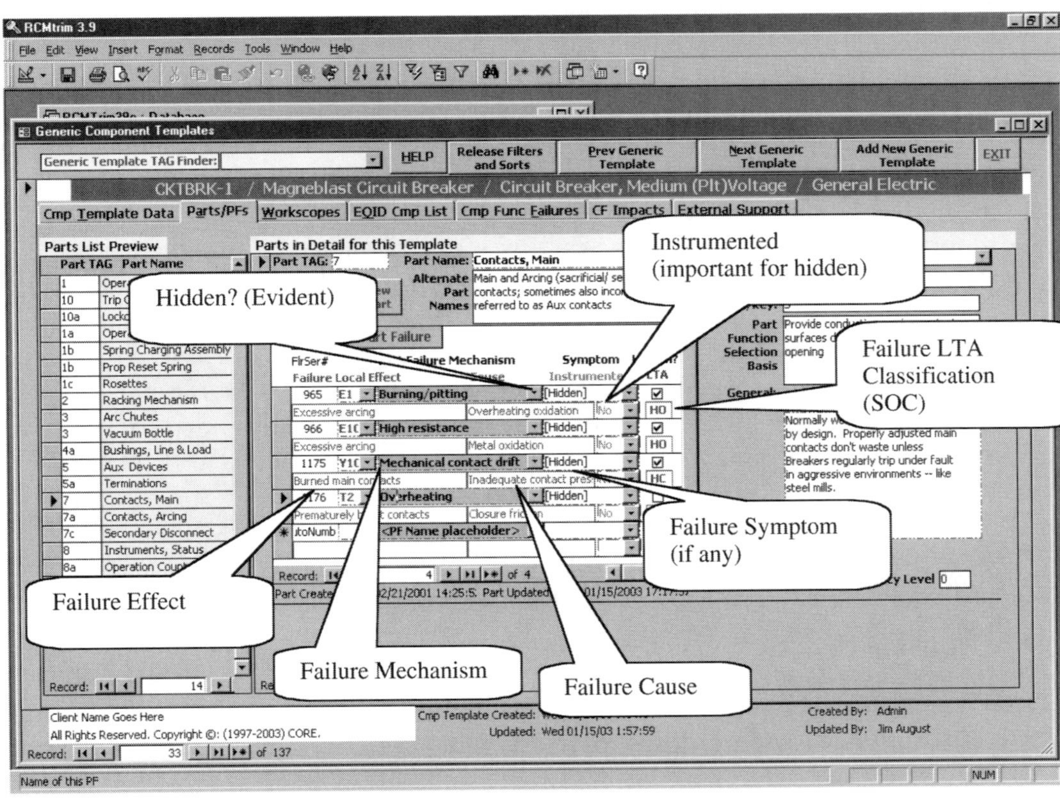

Fig. 5–8 Failure Causes and Local Effects

In many circles, equipment PM work is considered conservative for failure prevention. This stems from an assumption that maintenance is exact; when performed correctly, perfect outcomes result, and perfect maintenance performance is expected. This is a deterministic mindset, one that prevailed in regulatory oversight programs in the past and is still pervasive today. Practically, maintenance can also be viewed probabilistically. Experienced workers yield more probable outcomes, but any maintenance has some uncertainty. (Even the best workers screw up, just far less often. Some maintenance is inherently difficult to perform successfully.) The more complex the equipment, the more uncertain the outcome.

One of the classic conclusions of aircraft turbine maintenance studies from the early 1960s (RCM's development period) was that contrary to contemporary FAA regulatory presumption, maintenance performance on aged but not obviously degraded jet engines greatly increased future probability of failure! (Nowlan and Heap, *Reliability Centered Maintenance*). This study established infant mortality and uncontrollable randomness as natural considerations for any complex, overhauled equipment maintenance, regardless of mechanic skill level.

Nowlan and Heap introduced the inescapable conclusion that adopting reliability techniques improved equipment failure outcomes. With the effort focused on improved outcomes, the concurrent need is to understand equipment aging while conducting intrusive maintenance on apparently adequate equipment. The sole justification for intrusive maintenance without known failure is taking opportunity samples for age exploration in fleet-leader aging components.

Aging life

In contrast to complex equipment, some equipment or parts of equipment exhibit very specific failure modes that are predictable after a certain age and account for a high proportion of failures. Such is the case with electric motor failure. Statistics show that 45% of the time motor failure is caused by bearing failures. These in turn cause winding damage after the rotor center position air gap is lost, wiping the stator coils (which is secondary failure). Winding deterioration failures (22%) from aging occur after long periods (12–15 years) of continuous service for large, Class H high-voltage motors. The remaining failures can be attributed to a variety of causes that for practical purposes may be treated as random. The following are the key facts that these statistics point out:

- Motors as a group exhibit two dominant failure modes—bearing and windings failure.
- Most failures that require rewinding are due to bearing failure and consequent secondary winding damage.
- Eliminating bearing failures leaves winding aging failure as the next major failure class.

Motor failure modes—secondary failures excluded—are roughly split between random bearing failures and aging winding failures. Anti-friction bearings exhibit approximate random failure. Studies indicate that installing anti-friction bearings requires much skill and most failures are due to faulty installation. Failure can also a result from careless lubrication practices that inadvertently shorten the part life. Consider the implications for engineers trying to reduce bearing failures. (Perhaps this explains the trend to use factory-installed sealed bearings for the life of the equipment!)

Aging life failures are the *no-brainers* of failure. Their importance stems from certainty of failure when the age life is exceeded. Few things embarrass a conscientious reliability engineer like missing an obvious age-based failure in an equipment-failure analysis. Misses come from fundamentally misunderstanding a type of equipment, material, or service application. Cost-based parts substitutions by penny-saving buyers also cause failures. To save money, these cost-conscious purchasers throw specifications and supporting analysis away on critical parts like key belt drives and valve seats.

Purchasing departments are never held accountable for a plant's *going down* due to the failure of an inferior part, nor for high maintenance and rework cost due to inferior part quality. Corporate policies that allow unqualified reviewers to make substitutions for specified parts are the root cause. It's pointless for companies to invest in reliability but not fund the related quality programs required to deliver it.

When aging failure is discovered, it's an obvious opportunity to use simple, traditional preventive maintenance—time-based or hard-time tasks that rework or replace aged parts. Cleaning of turbine blades to remove steam deposits constitutes rework; replacing lubrication in oil-lubricated equipment such as induction draft (ID) fans constitutes *replace*. Rework today extends the lubrication oil by formerly inconceivable restoration techniques. Lube oil rework can include oil filtration, water removal, and additive replenishment. These techniques restore oils to like-new conditions.

Most often aging-based failure identification comes from application familiarity, engineering failure modes, and fundamental aging physics. On a fundamental level, parts and materials deteriorate very few ways. Becoming completely familiar with these processes and knowing the physics that cause them is part of learning equipment reliability. Materials science and equipment design classes in engineering curriculums help develop this knowledge base.

The challenge of aging-failure identification is to find appropriate aging parameters and monitoring techniques to quantify equipment age for hard-time, time-based, or on-condition monitoring-based replacement. Without a well-defined aging parameter, equipment aging dispersion makes hard-time replacement risky and less economical. The best strategy for controlling aging failure modes combines a known effective condition monitoring technique with a known aging parameter to bracket the expected failure condition onset. This strategy is economical for large machines such as turbines where it's necessary to control final wear-out modes that exhibit dispersion to exact intervals.

Variations in operations stress introduce aging dispersion. On significant rework expenditures like overhauls, exact timing makes control of dispersion effects (to forgo the use of hard-time maintenance) highly desirable.

On-condition techniques measure decreased resistance to wear and determine appropriate rework or replace conditions. Sampling lubrication oil in machines that are run intermittently without maintaining operating records provides an alternate method to determine their lube oil condition. Aging-measure selection may effectively allow hard-time maintenance where parameters like number of starts, tonnage, or integrated volume can be measured and tracked to reasonable time maintenance performance.

Random failure

Random failures are the opposite extreme from aging ones. Failure randomness reflects variation in stresses, complexity, or other factors. Randomness in failure is much more common than aging. One behavior-based study of failed aircraft parts found that 93% of failures are cited to be random or nearly random.

Implicit random-failing equipment models abound, although few failures are purely random. These models are simple mathematically, and yield representative results for system failure simulations. Their validity in models certifies random failure predominance. Random failing components include electronics, bulbs, and electrical devices like diodes and foil capacitors. For these, randomness stems from environmental stress variations such as voltage, moisture, and heat. Environmental stresses are highly variable, hard to control, and cause unpredictable and sudden electric and electronic circuit effects. Semiconductor breakdown can occur for many reasons; when thermal runaway occurs, failure is sudden and complete.

Mixed failure

Real-world failures mix aging and randomness. This distribution is modeled in its most general form by the Weibull distribution, which includes infant mortality (*see* Fig. 5–9).

A Weibull paper (or diagram) is like log-normal or log-log graph paper. It reduces a failure mode to a line and identifies key Weibull characteristics from the linear approximation when there's a good fit. With Weibull technique, goodness of fit can be determined visually.

Although most failures are mixed, aging predominance introduces hard-time failure control options. In the absence of aging, condition monitoring is needed to detect failure onset. Addressing randomness requires adoption of design strategies such as redundancy. Separating mixed data into distinguishable failure mode components may be accomplished with the use of analytical subroutines or Weibull paper. Once decomposed, multiple failure patterns may become evident that reduce the randomness of the remaining data and uncovers distinct failure mode contributors. This technique is advanced and is used only occasionally with statistically significant failure populations. Presented with mixed failure, analysts should look for data trends and treat analysis with random failure strategy controls.

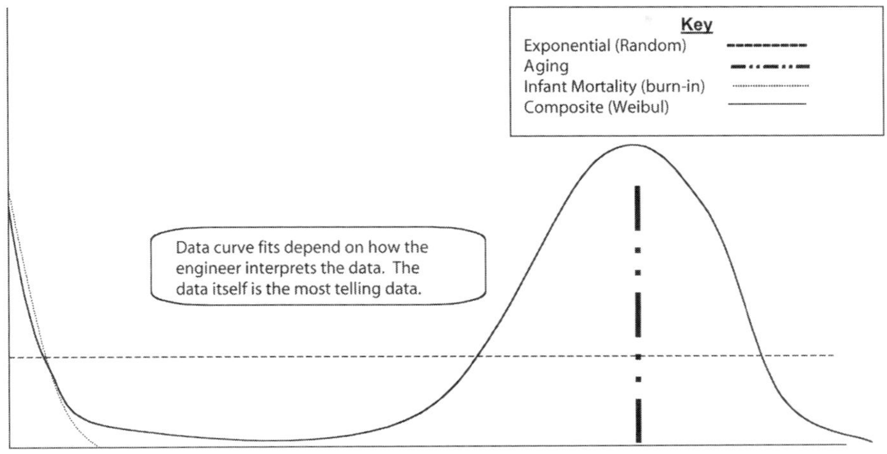

Fig. 5–9 Weibull Distribution

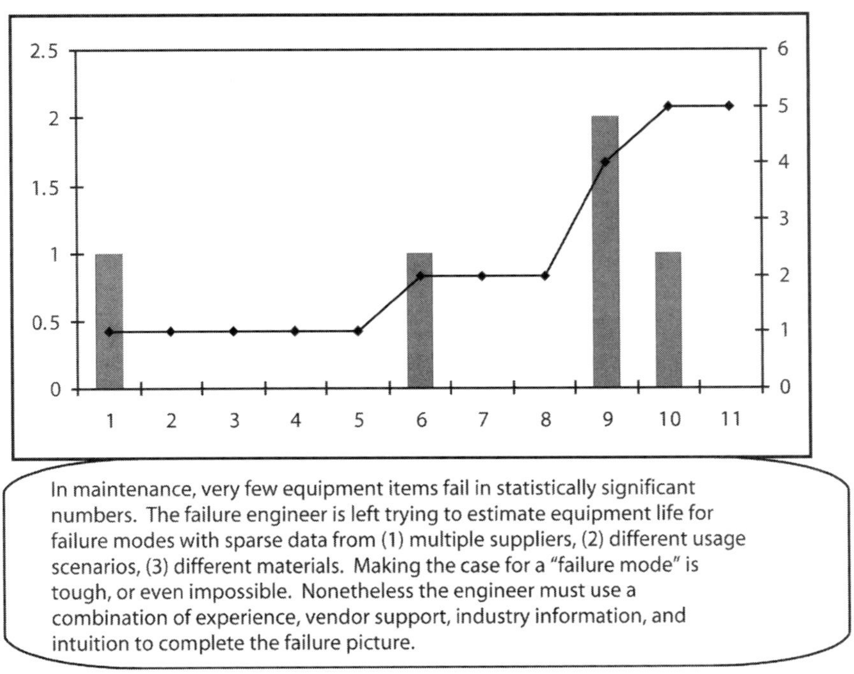

Fig. 5–9a Sparse Data Weibull Distribution

Estimating lifetime

Developing failure-aging data to estimate part life requires collecting part-aging data to build a failure distribution curve. Identifying failures requires reading WO failure reports. Although some WO systems have failure identification fields, reading as-found text description recorded by maintenance workers helps identify failures. Failure events can be summarized based on written descriptions into failure-types. Identifying these separately and adding new events as they are identified can quickly build a failure-frequency distribution. With this distribution, one can construct a Pareto chart—a bar chart ranking failures in order of frequency. Statistically representative failure samples are needed to clearly identify dominant failure modes requiring enough operating history coverage to represent operational use (*see* Fig. 5–10).

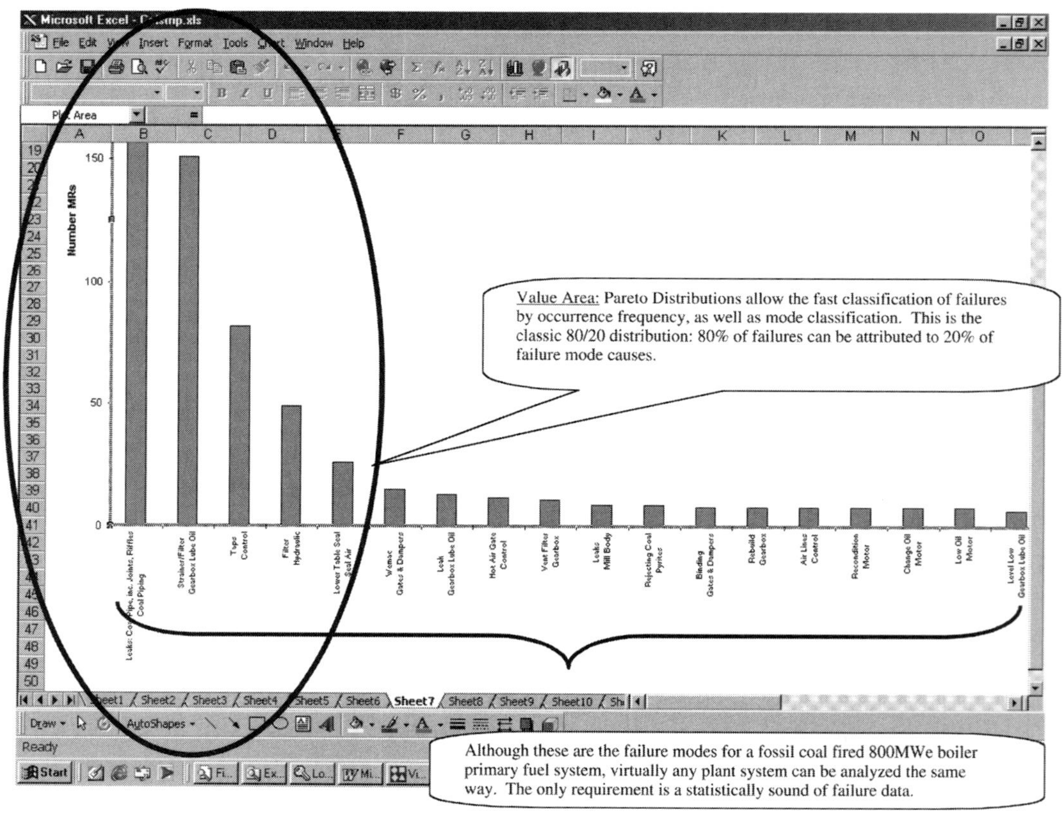

Fig. 5–10 Pareto Chart Example

Typically, 1000 corrective maintenance WO reviews are required as the thumb rule for any particular class of equipment or system. Some plants will not have that many equipment WOs in their entire history, and that points to another problem: working with sparse information. In practice, after examining a large population, the developer gets a feel for when the distribution shape has matured. Like watching a Fast Fourier Transform (FFT) develop from analog data on a computer screen, as the volume of failure data reviewed reaches maturity, no new failure modes are learned. As new failure-mode encounters decline, the profile becomes statistically mature and all numbers increase uniformly. The distribution shape grows, but doesn't change proportionately.

Using WOs review to estimate equipment's age lifetime requires representative data. Too often, the data provided is filtered accidentally and is not statistically representative. Methods for sampling data to assure statistically representative populations are subjects left to statistics books and the design of experiments. Needless to say, this is not assured without forethought, planning and experience. Once representative WOs have been selected, the age estimate of equipment at failure can be made. The failure distribution itself will exhibit the point at which aging life becomes apparent. When unclear, tools such as Weibull analysis or statistical equivalents may be brought into play.

Determining part-life information is often the single most qualitative, numerical estimate in RCM. In the power industry, part-service histories—when a part entered service and when it was removed— usually aren't available. The parts stocking system may provide an alternate data source to WO CMMS systems for estimating service life. First date of purchase helps indicate service lives for large, one-of-a-kind equipment and parts. Workers engaged in problems on equipment that requires repetitive repair will readily talk about parts quality and usage on that equipment. This provides another way to validate parts service life.

Workers know approximate part service lives whether or not they record lifetimes on WOs. Plant engineers dealing with failures should sensitize workers to the need for part aging data and its practical use. Developing a culture of failure analysis is a major step building a living maintenance program. When statistics aren't available, surveying those working on the equipment is often helpful. Although inexact, surveys are usually acceptable for cost analysis and cost-based failures, which cover many common analytical cases.

Benchmarking with other plants is another useful technique. Industry failure-engineer professional groups, manufacturing associations, and other trade associations are helpful. Although they have a bias, manufacturer representatives are also very helpful. They have access to parts information—how often others purchase replacement parts, upgrades, and so forth—that indicates service life. Manufacturer engineering groups also can be a source of information. Members of these groups share opinions and estimates off-the-cuff with their peers more openly than the supplier's sales organizations. For high-dollar cost failures, insurers and groups like EPRI often pursue initiatives to control losses or improve equipment service; therefore, those organizations can provide much needed data. Where safety or operational impact occurs, exact statistical information gains even more value.

Codes address many safety-related part lifetimes. Safety valves must be lift tested quarterly, checked for liftoff setpoint every 18 months, and rebuilt every 36 months as one example. Safety-related part lifetimes not controlled by code usually address control and alarm loops where integration has obscured the safety functions. Extending intervals from the conservative limits already established by existing codes requires solid statistical data and even new code cases with the regulator and/or governing code body. Changing code limits does not directly result from reliability study, although indirect benefits are impossible to predict.

Statistics developed as failure density histograms of raw data, developed around single failure modes, tell the failure story too. Focusing on singular-failure modes is the key qualifier. Engineering tests such as Weibull distributions, either on plotting paper or software, help confirm single-failure mode predominance. Where two or more modes are present, they are evident as two predominant line/slopes on Weibull paper.

Failures follow many distribution patterns. Most failures are *random* or exhibit highly random aspects. Extraordinary aging patterns are special exceptions. Analysts look for trends to confirm patterns and analytical models. The failure distribution itself can reflect anything—even a composite failure mode.

Based on failure mode, confidence in the aging information, and its effects on operations in a safety, production, or cost context, analysts can decide how to proceed to develop a control strategy. Failure modes affecting operations are the toughest to identify because they're ranked not *safety* but above *cost*. Cost is a modest concern in generation. Safety modes incur major concern everywhere in industry today. Analysts have some latitude dealing with operational failure; however, they must be aware that operational failure carries significant production risk. Conservatism is taken for granted, based on large costs for production losses and data sensitivity in plant RCM analysis. Production losses invariably outweigh simple cost-based ones by orders of magnitude.

A plant sought to determine how to modify generator overhauls based on information related to several catastrophic generator winding events. One case involved an extremely expensive operational loss. Stress cracking should never have occurred, according to material selection. (Stress cracks require prolonged exposure to moisture.) One prevention option was a costly, high-risk inspection for stress cracking conditions on super-alloy steel rotor retaining rings. At issue was whether inspections would have any value or introduce more potential for damage and whether at-risk conditions existed on any other units. The discussion included the insurer who had covered the loss. With a desire to avoid future losses, insurer and insured asked, "How do we avoid future losses?" The question ultimately rested on the nature of the failure itself: was it a dominant mode, or a rare unmanageable event?

One answer is to replace any part with a lifetime in advance of failure with 100% certainty. For safety equipment, the single most important outcome from failure analysis/actuarial life studies is that direct safety failure modes can have safe life limits. When condition assessment or tests come into play, they must predict the PF interval, removing the risk of the failure occurring prior to rework. Boiler pressure welds (steam drum), reactor welds, rotating mass part cracks, corrosion mechanisms—all are examples of mechanisms with direct safety implications. Where industry experience is available, searching for similar industry applications based on reported disasters are imperative.

As an example, plant analysts inspected high-energy pipe bending locations for corrosion erosion wall thinning—based on configuration and low-oxygen erosion/corrosion potential similar to a Surrey Station high energy pipe that failed 15 years ago. The generating plants inspected were fossil-fired, so no direct requirement was imposed by a regulatory body. However, a number of potential *at-risk* areas (based upon low dissolved O_2) were identified. Upon performing UT in the susceptible areas, at least one obvious incipient failure was found, leading to a direct save.

So, developing failure statistics reduces to completing a partially painted picture with judgment. Some engineers do this well; others do not. In the final analysis, the exercise comes down to confidence and judgment. In all but a handful of cases, failure issues surrounding direct failures are one step removed from the process of identifying dominant failure modes and classifying failures as direct or secondary.

Developing Failure Statistics

Industry statistics

Industry statistics are available through trade groups and industry organizations. Quasi-regulatory bodies in North America, like INPO and NERC also fit the latter role. NERC collects and disseminates failure data for generation and transmission operations in North America. The NERC reliability database and FERC cost database maintained for regulatory purposes nonetheless provide a ready source of benchmark and specific failure data. Rules and agreements ensure utility participation. Some companies' NERC statistics are more useful for reliability purposes than those they internally prepared! NERC statistics are excellent reliability study source material on similar plant groups at the system level. They readily support benchmark comparisons against anonymously identified NERC-equivalent benchmark plant performance groups.

FERC reporting rules require that plants develop and submit reliability and cost data. FERC cost-reporting categories were established 30 years ago, however, and reporting categories (before reliability ideas had advanced) limit data value. These legacy reporting areas reflect plant physical layout rather than systems. Regularly working with FERC data, however, analysts learn to interpret these obsolete area-based categories.

Using internal data, reporting parties (plant staffs who report unit unavailability events and causes and/or failure/cost accounting support groups) haven't always been sensitive to the need to report data accurately. Classification, correctness, and detailed exactness have historically been low. This made overall data confidence low. Renewed emphasis on submitted data validity as a result of deregulation has resulted in initiatives to improve reporting. Underway for five years at this time, they look very promising. Recent Northeast blackout events are likely to increase that focus and trend on accuracy of NERC failure reported data.

Site statistics

Historically, generating plants weren't concerned with performance statistics. Only the past decade's quality and competitive awareness experiences have made site technical support and plant managers more aware of value of data and its uses for performance measurement and improvement. An old saying goes, "What you measure improves."

New software systems introduced over the past decade promise to redress some historical utility information problems. New CMMS/EAMS systems should have better parts usage information capabilities to support aging studies. CMMS/EAMS failure identification, classification, and reporting user friendliness have substantially improved.

These latter fields' entry requirements have vexed reliability engineers, planners, and occasional users (operations) over the past 20 years—spanning two generations of CMMS. Early systems required code lookup and entry to close out work orders. Maintenance supervisors won that lottery and took on that responsibility to enter failure codes on complete WOs. When managers required worker coding entry to approve time cards, users found codes that the cost accounting CMMS systems would accept and used them to obtain approvals so they would get paid. Users work around unfriendly systems, and work goes on. Floor personnel are uncanny about sensing motives and working around them. Site part usage and failure statistics collected over the past 20 years are still highly suspect and may be for 20 more years. Developing ways to get useful data remains a great daily reliability engineering challenge.

Inference

From the context of work orders, an experienced engineer can infer many things about both the work done and the reason it was done that aren't explicitly documented. For example, many plants have policies to replace parts that influence routine parts usage. Mandatory parts replacements could be inferred to be unusable failed parts, for example. Using parts replacements from stocking to estimate part failures, making this assumption, would greatly over-count the number of parts failures in service. Consumable parts are simply replaced. These include many nuts, bolts, gaskets, seals, O-rings, and other normally non-reusable/reused parts, as well as serviceable parts. Types of repairs and repair duration can be inferred from the duration of outages and work order times.

Boiler tube pad welds take much less time to make and support than do quality window welds. Repetitive repairs typically indicate recurring problems. Calibrations are repetitive and planned. When calibration results fall outside acceptance criteria, they constitute failures. Failed calibrations are repetitive in some plant venues. The challenge in some instances is inadequate design; in others, intervals exceed the equipment capacity. Although the long-term resolution of many calibration problems is improved design, the adjustment of intervals is needed to address repetitive, short interval drift. A more frequent problem in calibration programs is the inability to extend intervals on equipment that lacks drift. In the extreme case where instruments exhibit virtually no drift, calibration may be unnecessary. Many instruments that would be classified *non-critical* based on function receive calibration. The potential work reductions in calibration programs from systematic risk expression classification, followed by age exploration, are substantial (*see* Fig. 5–11a & b).

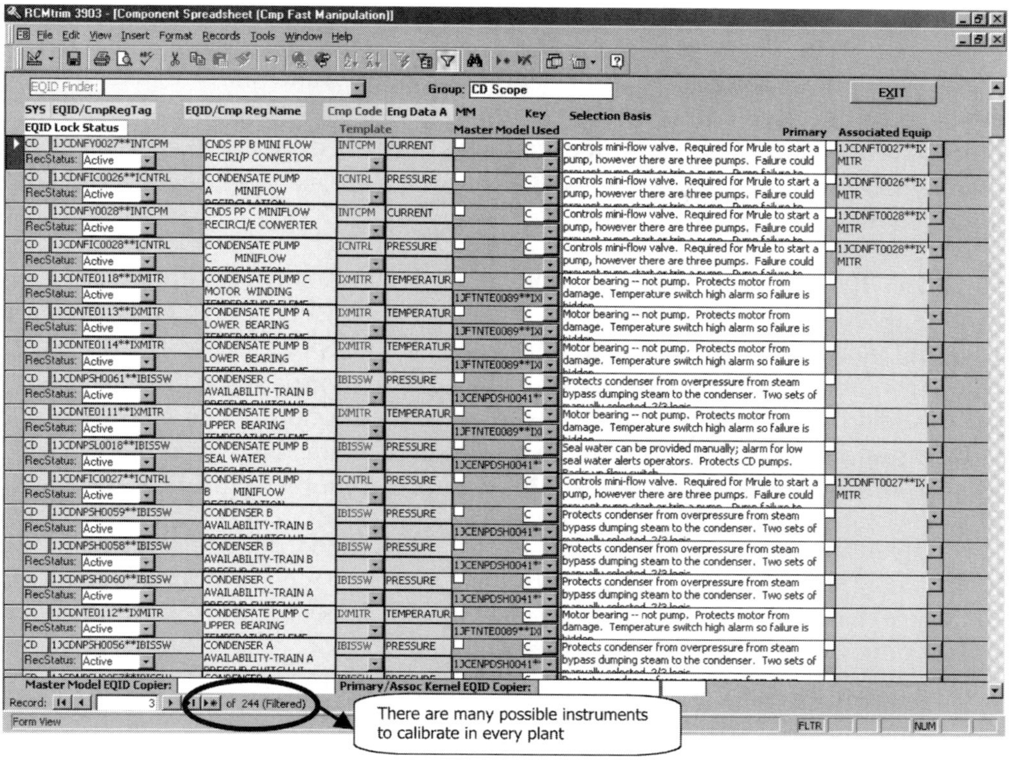

Fig. 5–11a All Instrument Calibrations

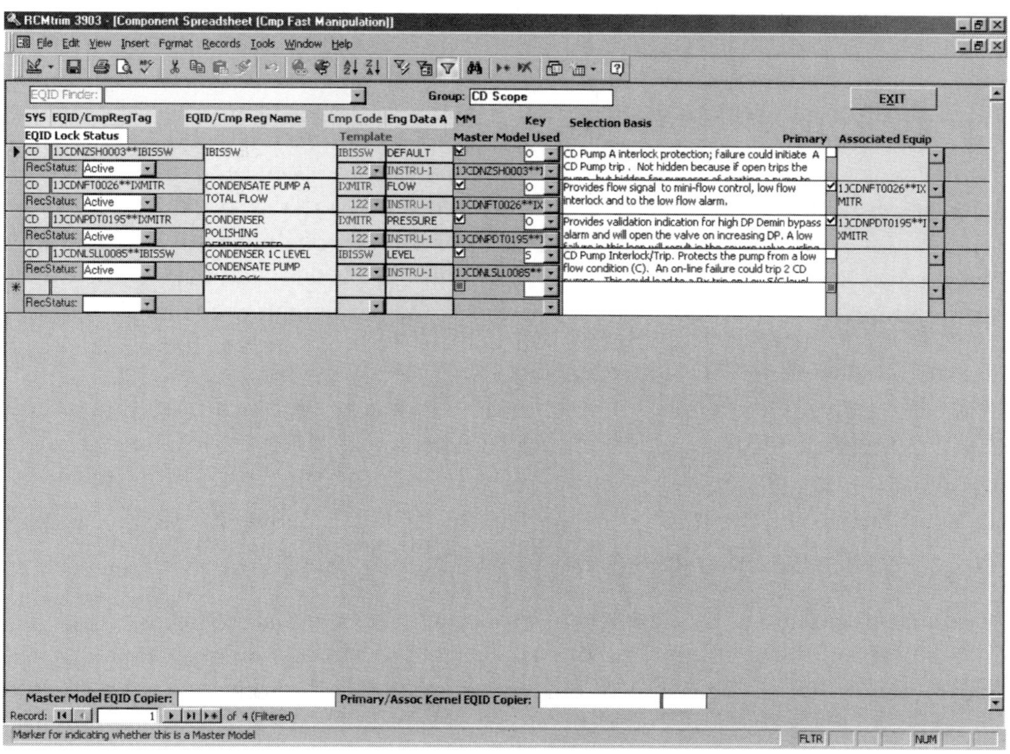

Fig. 5-11b Critical Instrument Calibrations

Leading age samples

Every fleet has its most rapidly aging components and parts that can provide leading age samples in a conscious design program of age exploration. Where safety or operations are at risk, these age-studies may be prescribed by licenses. Nuclear plants commonly have metallurgical coupons that must be removed and examined to validate the material aging processes.

Aging samples can be used to validate assumptions about part life in cases where they aren't formally required by licenses. Sometimes it is just good reliability engineering to use aging samples. For example, for one company aging analysis on Stellite hard valve seats in a coal unit's pulverizer primary air shutoff valves provided valuable insights on the relative merits of hardened valve seats, reworked seats, and discounted seats. While most plants and engineers perform these studies as a matter of course, the challenge is to add the knowledge gained to the corporate information repository.

Every plant needs a program for age exploration. Many times engineers fail to use documented aging studies to extend PM task performance intervals simply because they aren't aware that it is a natural consequence of a living program, or are unaware that it is plant policy, as well as failure engineering practice.

Hidden failure and redundancy

Hidden failures are at once simple and mystifying in RCM analysis. The term *hidden failure* is a direct outcome of RCM development. The revelation that some failures are evident to the operating crew is both obvious and filled with mystical insight. In fact, the purpose of hidden failure identification is to ensure that important failure modes do not remain hidden.

Historically, everyone in plant support becomes aware from time to time that certain failures were hidden. The value of the RCM approach is that if organizations embrace an RCM philosophy, the hope-for-the-best approach to dealing with hidden failure is superseded by a rational, careful strategy for control. Organizationally, this approach provides a place in operations' work list to check the many high-risk alarms and standby items that, historically, operations just hoped worked in a real demand event.

Hidden failures are simply failures inside the operator's black box. Open the box, and failures appear. Most hidden failure maintenance strategies lead inside the box. Literally opening a check valve's internals to look for the presence of a hidden failure (let's say a loose hinge pin) goes inside the box for an inspection. A periodic function test to validate that the valve prevents backflow might do the same thing, non-intrusively. Verifying functionality may validate that the failure hasn't occurred, depending on the design and analysis. Performing the work, in either case, reveals the hidden failure. Taking equipment out of service for intrusive inspection can always reveal hidden failures. The objective is to do identify hidden failures in other, better, less-expensive ways (*see* Fig. 2–15).

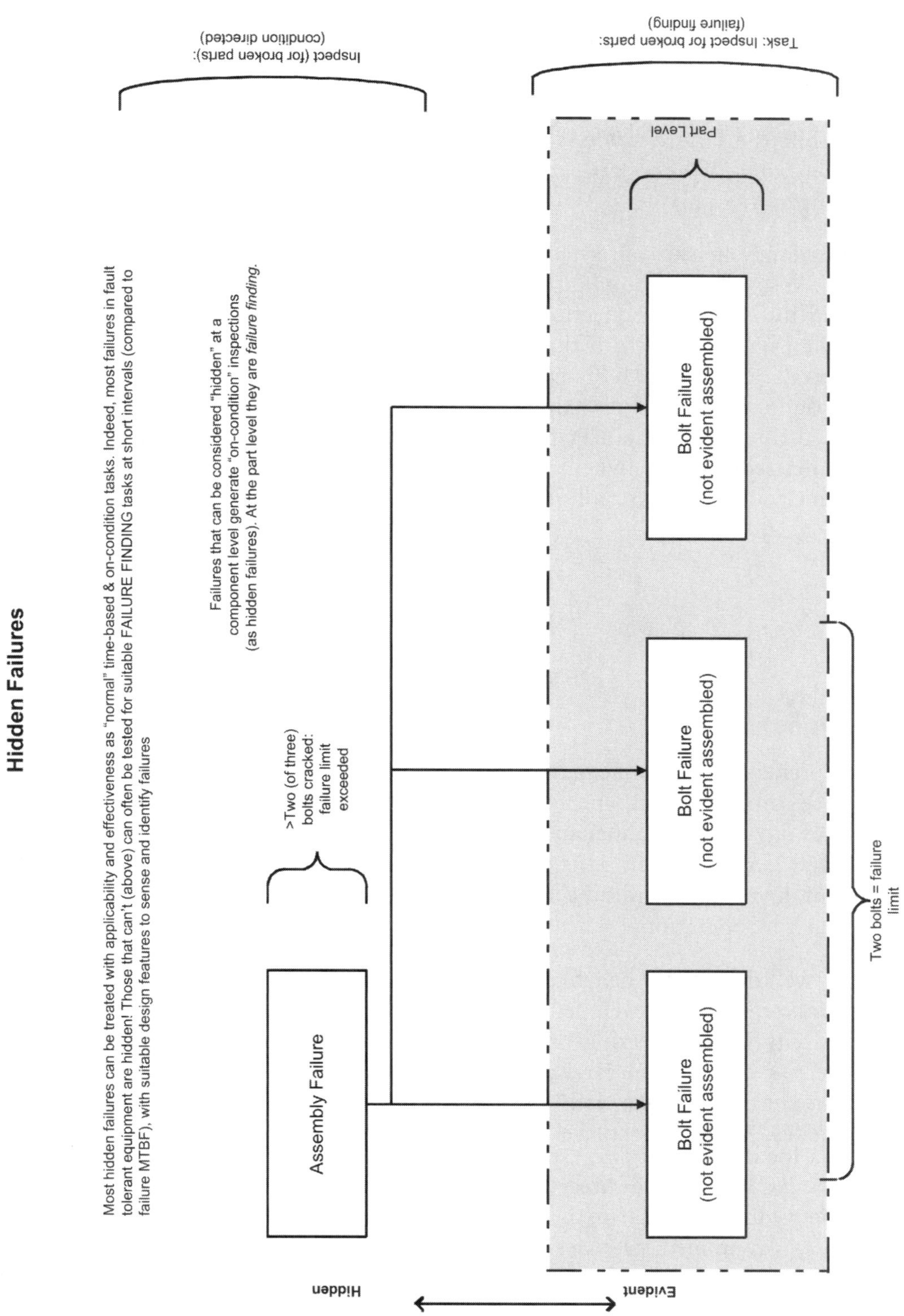

Fig. 5–12 Hidden Failures

Many high-risk hidden failures are instrumented to make the failure evident. For example, a low lube oil pressure sensor and alarm on a critical bearing alerts the loss of pressure that would lead quickly to bearing failure. The combined bearing failure—on low oil supply and instrument-pressure alarm circuits, form a part failure-instrument pair. Without the part failure mode the instrument has no purpose or value. While the failure mode is absent, the alarm is redundant; the alarm operates only upon failure. These pairs are called equipment-instrument failure pairs, or just equipment failure pairs

Redundancy masks paired equipment capability loss. Redundancy strategy is to reveal the loss of the redundant counterpart. Service equipment losses that become embedded into plant design effectively redesign the plant. This occurs when failed equipment isn't restored in a timely way. Paired function-providing equipment and instruments are common. Reliability engineering addresses hidden failure through the most cost-efficient techniques. Many hidden failures exhibit aging and can be treated with hard-time maintenance tasks. Providing condition-directed maintenance performance reduces costs when intrusion is required, an on-condition task is effective, maintenance cost is high, or all of the above (*see* Fig. 5–12).

Risk Exposure

SOC distribution

Comprehensive component risk exposure ranking provides an operator reference for relative component risk. Incredible events—earthquakes, explosions, terrorists—are not part of day-to-day maintenance focus. Maintenance deals with failures incurred by design and routine operation of the plant. Leaky components, balky standby equipment, line cracks, and tube leaks are common maintenance occurrences that must be managed by operational schedules.

One well-developed benefit of risk exposure partition is *relative equipment differentiation by risk*. Developed well, the partition provides an immediate operating risk hierarchy for engineering or management, an 80/20-risk road map. Each level down the risk exposure hierarchy (SOCX) reduces operating risk. A correspondingly larger amount of equipment falls into progressively lower risk categories for dominant failure modes. At the lowest level, there is no risk based on non-critical characteristic.

Airline RCM provided direct safety risk restrictions for safety risk classification. A direct safety-failure risk has immediate consequences for the operating safety crew and passengers. An immediate safety risk exceeds all other work barriers. Excluding indirect safety, risks may seem non-conservative, but many risks can be immediately removed by ending a mission. For example, a three-over redundant hydraulic control system can suffer two independent failures and still offer hydraulic control. For an airplane, loss of hydraulic control is an immediate safety risk, so redundancy is provided. Independent line routing, reducing common-cause secondary failure risks from rotating part failure *missiles*, eliminates common cause events. One hydraulic system is the minimum needed to operate a plane. Because one is insufficient to manage risk of loss without jeopardizing safety, the existence of only one functioning hydraulic system immediately terminates the operating mission. The plane must land at the first and easiest opportunity. In commercial airline RCM, critical usage is restricted exclusively to safety failures.

Where three systems are available, one hydraulic train loss leaves one primary and one backup subsystem. Backup remains, but further loss affects safety. This situation would not terminate the mission immediately, but it would preclude beginning a new mission until the equipment is restored to the design-basis configuration—three independent operating hydraulic systems for control—are completed.

Aerospace RCM introduced a basic idea: direct safety risk with redundant layers and dynamic risk shifting based on the following layers of defense:

- safety: minimum one, operationally capable two
- operations: minimum two, mission commencement requirement three
- cost: minimum three

Redundancy shifts risk exposure down. Adding one level of redundancy changes an operating risk to a cost risk by removing the operational impact of a failure. One additional safety redundancy layer shifts a safety risk down to operational. Redundancy provides operating leverage.

Management policies should use redundancy to their advantage, but redundancy use policy must be clear. In some operating situations, policy is muddy, obscure, or confusing. Confusion complicates application, thus introducing risk. Grace periods allow condition workarounds while condition-directed maintenance is performed. RCM philosophy emphasizes necessary on-condition maintenance performance; the concept can be a cultural shift for operations that are in a risk-ignorant, run-to-failure mindset. Policies for the performance of online maintenance must also be clear in advance. Operators should (but don't always) appreciate failed equipment operating risks; they may learn to operate with failed equipment risks indefinitely.

The case of an underground storage facility demonstrates this point. Failed methane detection alarms for PRB underground coal storage removed the hidden failure risk of undetected high methane concentration. This detection feature was provided by design. Failing to restore methane detectors to service in a timely way violates on-condition maintenance performance assumptions. (They were checked, found failed, and should have been restored within a brief period.) Once failed, the operating risk of methane accumulation and explosion is unprotected. Hidden detector failure introduces a hidden safety risk originally neutralized by the original designs' inclusion of the detection mechanism. Though methane accumulation is infrequent, such cases occur. In most cases, loss of safety alarms reflected risk-awareness ignorance, not intentional safety violation. RCM analysis reveals their risk, placing it back into proper perspective.

Critical equipment risk ranking is stratified; only a small part of the system's equipment has a safety classification. Even in safety systems, it's unusual to find more than 50% of the equipment with direct safety failure potential. Redundancy is the reason. Because of their safety importance, these systems invariably have high degrees of redundancy. So, even though failures are direct, consequences are minor. In nuclear plants, for example, safety systems have fully operational standby systems: two or more fully independent trains. The failure of any one standby train forces the plant into an action statement that requires redundancy restored within a period of 24 to 72 hours, after which the plant must be shut down. These conditions reflect aerospace's MSG-3 standard—reduction in safety margins terminate the operating mission.

Equipment SOC categories' relative proportions vary system-by-system. Some systems have higher or lower ratios, so understanding *why* is useful for comparison. Comparisons allow benchmarking the classification process, discovery of system risk profiles, and other analytical activities. Not surprisingly, safety-coded components are found in non-safety systems, and vice versa.

Equipment rank assignment presents a puzzle: No component, train, or system has a risk classification based upon its hardware name or physical configuration. Function and failure drive risk. Functions, their loss, and failure modes rank equipment risk. Equipment classifications help simplify and develop risk categories—albeit artificially. In the final analysis, equipment roles and names mean very little without failure mechanism's context to develop operating risk.

For example, history shows that bleeder trip valves remove steam backflow potential, among other functions. Steam backflow has a safety risk of overspeeding a tripped turbine causing ejection of missiles. That risk, in turn, introduces a safety-based bleeder trip-valve failure risk. Yet, the extraction line risk is based on cost: a failure to extract steam causes efficiency losses and increases costs. (This latter risk ignores the passive extractive line function to contain the extraction steam—a passive structural function. The loss of integrity function for extraction steam is not a credible DFM.)

It is incorrect to think a system's equipment risk exposure rank drives system risk ranking. Failure mechanisms really drive risk, based on system requirements. Plan analysts should be cautious if they encounter *any* equipment risk classification without failure modes. A component that can't credibly fail causing a safety, operational, or cost functional failure should be ranked non-critical. A component that can't credibly fail, causing a safety functional failure can be ranked as non-safety. A 10-function failure component with one failure mechanism that causes a safety functional failure is classified critical, but the 9 others may not. Allowing the single safety failure to drive the balance of failures upward would immediately lead to heavy maintenance programs.

In fact, to classify the other nine failure mechanisms with a higher risk on the basis of equipment classification based in turn on the first safety mode, inappropriately overvalues the nine. This analysis is incorrect, and incorrect analysis frequently overvalues many activities. Other work must then compete with overvalued work, which clouds the focus on legitimate safety or operational failures. Erroneous ranking confuses rather than clarifies. Again: *failure mechanisms drive risk*!

When encountering composite equipment, trains, or systems assigned single-risk classifications, analysts should consider likely outcomes of partitioning. Reduced to direct safety terms, few components' legitimate safety risk classifications are not already addressed explicitly by codes, laws, and operating standards. This strategy supports the subtle case for developing and maintaining an explicit basis. The explicit basis provides a high level of assurance that no safety failure mechanisms or an associated safety-failure mitigation task is ever lost. The net effect is a many-fold reduction in WO tasks that are based upon safety.

Excluded middle

Risk classification schemes explained by Nowlan and Heap document and rank (in MSG-3 [V2]) three risk levels based upon direct safety, operations, and cost category consequences. This scheme ranks tasks' risk hierarchy based on a general, broadly accepted differentiation scheme. Ad hoc risk classifications commonly used in practice are often problematic based on risk elevation introduced by other schemes.

As a general rule, most standby safety systems have at least one fully redundant train. Often these are provided in the form of an identical train. Using either MSG-3 or technical specification action statements, failure of these safety trains during plant operations—rendering the system inoperable—has mandatory production consequences. If the operating staff can't correct the failure inside the grace period, it must initiate plant shutdown. This is a classic operational impact.

An example of this type of operational impact involves a safety injection (SI) in a pressurized water reactor (PWR). SI provides reactor cooling, makeup, and backup shutdown capability in the event of several design-basis accidents, such as a design line break. Due to safety considerations, SI must be available at all times when the reactor is operating. Design provides two 100% capable SI trains in place to accommodate a line break in one of the trains. The second SI train provides redundant backup for the operating event. However, loss of either SI train puts the plant in a 72-hour *shutdown* action statement. Loss of the second train requires immediate plant shutdown trip. The first failure terminates the current operational evolution in an orderly way; the second requires immediate action. Under no case would a plant start up with an inoperable train. These nuclear action statements and their interpretations reflect the practical implementation of the guidance of aerospace MSG-3.

The fully redundant SI train provided by design makes the loss of protection layer for the design basis safety event an operating event under normal circumstances. Loss of the remaining SI train requires immediate production termination. Inability to restore a train in the allowed grace period results in an immediate orderly shutdown. Redundancy removes immediate safety implications from train loss, although operational consequences remain. Using the original MSG-3 RCM risk management philosophy, a failure causing SI train loss would be ranked *operational*. There is no direct safety consequence. A plant with only one SI train available would initiate an orderly shutdown.

Many lesser systems, like turbine and condensate, have turbine trips and reactor runback potential. Nuclear plants invariably rank component core damage scenario risk factors in safety terms. Doing so effectively considers multiple failures and removes aeronautical RCM direct-safety qualifiers. Any safety-classed system failure is ranked *safety* and treated accordingly. The nuclear culture's safety approach is imbued in the industry regardless of the practical operational impacts in plant operating license action statements. Even as the NRC's high-level policy becomes more risk-oriented—and less prescriptive—that change in philosophy hasn't worked outwards; plant sites remain in a state found 20 years ago.

Inadequate classification schemes elevate failure mechanism criticality. Operational, cost, or even no-direct-failure impacts are treated on par with safety. This approach dilutes the impact of truly risk-significant safety failure mitigating tasks that influence operating safety. Improving critical risk exposure review methods could balance this bias. Some authors point out that although EPRI introduced RCM nuclear plant pilots 20 years ago, nuclear generation never fully embraced RCM as a legitimate risk management technology in the United States. (At least one foreign-operating agency, Électricité de France, on the other hand, has endorsed RCM.)

The dilemma is how to rank equipment failures that affect operations but do not jeopardize safety. Any operational load reduction counts against safety system performance (10CFR50.65). These affect maintenance rule-reporting criteria, which are treated as safety events according to regulations. Nuclear operators, as a result, practically have no mid-range operations/production risk exposure partition classification.

Nuclear equipment is either S, C, or non-critical (X). The scheduled maintenance ranking of MSG-3's logic tree, so beneficial to classical RCM with its convenient three-tiered risk levels and relative work breakdown percentages, is lost. The consequences are that nuclear plants do much more work ranked at a much higher risk than would otherwise be the case. Performing all the required work drives nuclear maintenance costs up. Twenty years after EPRI's first efforts to introduce RCM, nuclear power remains fertile ground for further RCM development and application.

Ironically, the unintended effect of blanket assignment of work to the highest safety risk category is that the safety risk focus is reduced. At the same time, other non-consequential work stands side by side with work that truly matters. This exemplifies the Law of Unintended Consequences ("action taken could yield unexpected results") and is one possible outcome of ill-devised risk ranking schemes.

Workscopes

6

What is a Workscope?

Practically, workers manage many tasks on a single equipment work order. Planners organize equipment work into coherent, organized packages to facilitate craft's work. In evaluating prepared packages, well-developed WOs perform many different equipment failure mode PM tasks at one time. Multiple-task performance on a WO job establishes the utility of workscope (*see* Fig. 6–1).

The term *workscope* is borrowed from project management terminology that describes the scope of work in a project activity or an activity schedule. *Scope* is the scheduled activity's details specifically broken out specific for cost, completion criteria, and resources. A workscope assembles PM tasks for concurrent performance. Organizing tasks into workscopes eases implementation. Tasks consolidated into workscopes for work orders present fewer station work-order system demands (*see* Fig. 6–2).

Many plants assign senior personnel familiar with work practices to develop (or *block*) PM tasks into organized, easy-to-implement workscope packages. Performing groups of PM tasks more or less at the same time conserves resources and improves maintenance operating efficiency. One tagout boundary, WO, and scope define the WO like a complete project.

The case for workscopes

Planners organize maintenance work into efficient, performable packages. Workscopes allow planners to develop scoped work packages, based upon engineering-analyzed equipment failures. Engineers evaluate likely failures (dominant failure modes) and develop their mitigation strategy as part of the PM development process. This action defines work—establishing scope. Planners determine how best to organize

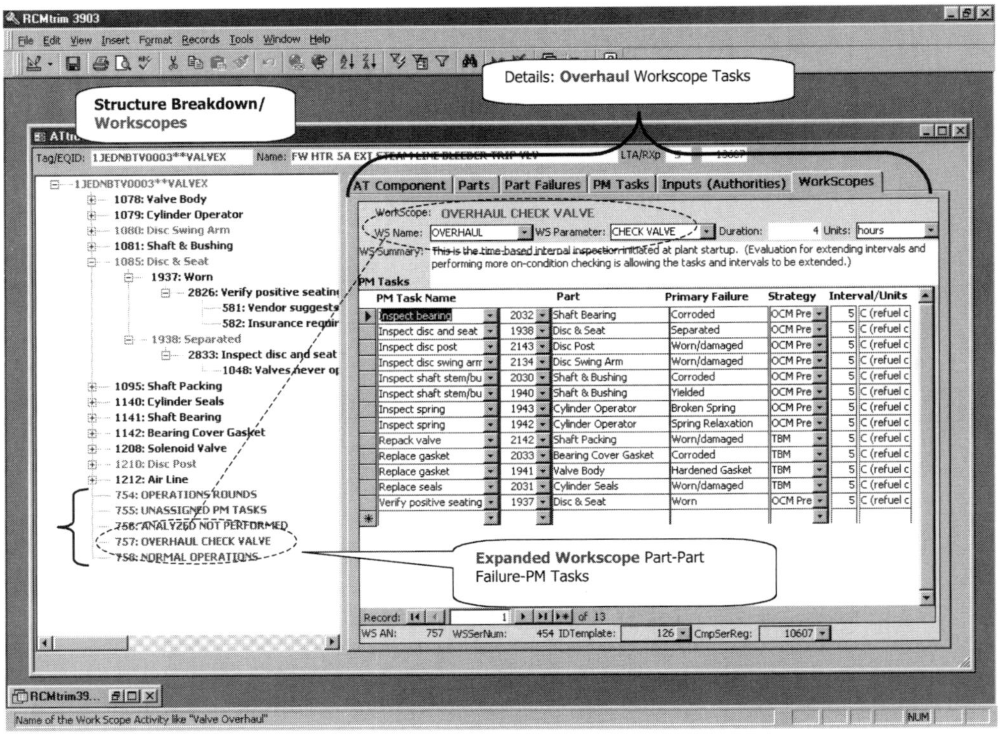

Fig. 6–1 Extraction Valve Overhaul Workscope

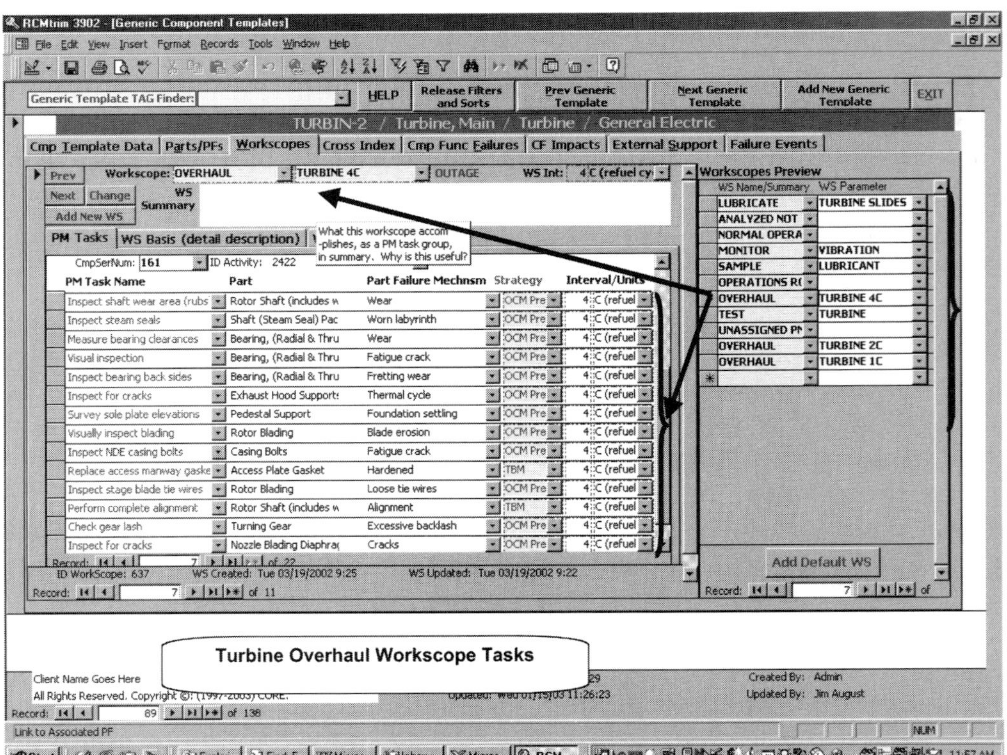

Fig. 6–2 Turbine Workscopes

the work for efficient performance. Enabling planners to plan and re-plan work orders, providing workers with delineated scopes to define completion criteria independent of engineering analysis, facilitates separate work development, implementation planning, and performance (*see* Fig. 6–3).

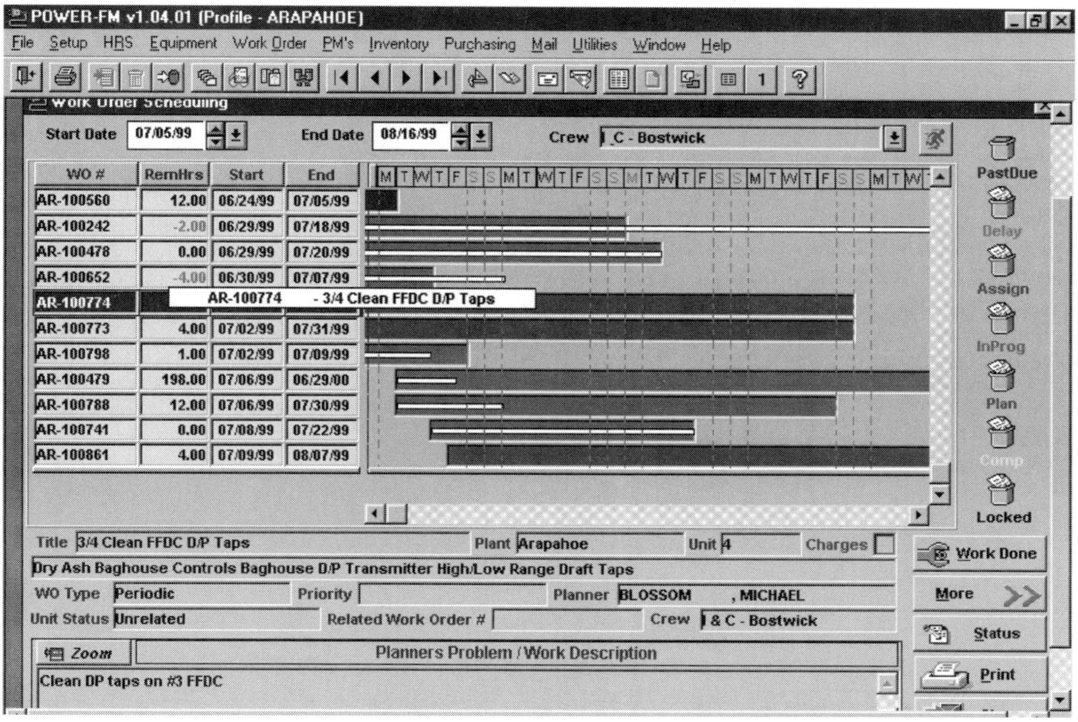

Fig. 6–3 Daily Work Performance Tracking

Practically, the development of high-value PM tasks proceeds along two independent paths. First, engineering develops the useful work that needs to be done to avoid failures using RCM-based processes. Planning the work and organizing it for efficient implementation follows. As each step occurs independently, there's an ebb and flow of tasks and their organization as new tasks and methods of PM performance go into production for testing and optimization. Depending on condition and technology, it may be easier to perform a predictive task (online) or a preventive time-based task off-line. It's a rare scheduled maintenance WO that's perfect on the first performance. In a dynamic world with ever-changing technology, even the best WO workscopes evolve as needs change. Methods to tune WOs need to be available (see Fig. 6–4).

Software workscope requirements

Grouping tasks into workscopes must be an easy process. Software subroutines should efficiently allow blocking and re-blocking of tasks to support workscope development. For an automobile, this would be like shifting "Check brake pads" from

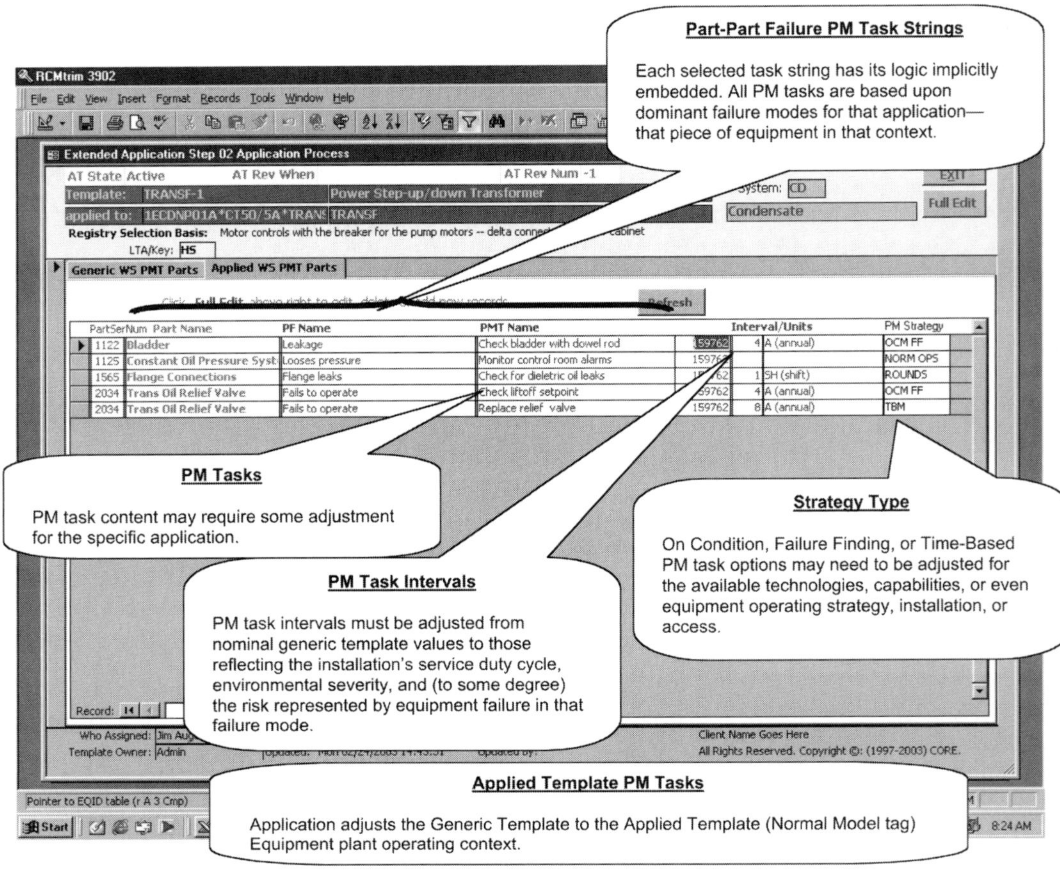

Fig. 6–4 Generic Template Workscope Task Edit

the 12,000-mile to 24,000-mile interval. During RCM projects, as craft worker and workgroup participation increases, the need to flexibly reorganize work becomes compelling. Point-and-click techniques to reassign workscope tasks automate workscope editing (see Fig. 6–5).

Workscope Performance Time Roll-up

PM time accounting

Scheduled maintenance requires less time than all other maintenance work. Excluding emerging condition-directed work, workscopes are known exactly; and work is either non-intrusive or well defined before it begins. Grouping PM tasks into large workscope blocks for easy performance is the most complex aspect in performing scheduled maintenance. Because task combinations that are most convenient to perform vary, the workscope becomes a useful tool for separating engineering task development from planning work and blocking it into workscopes by WO.

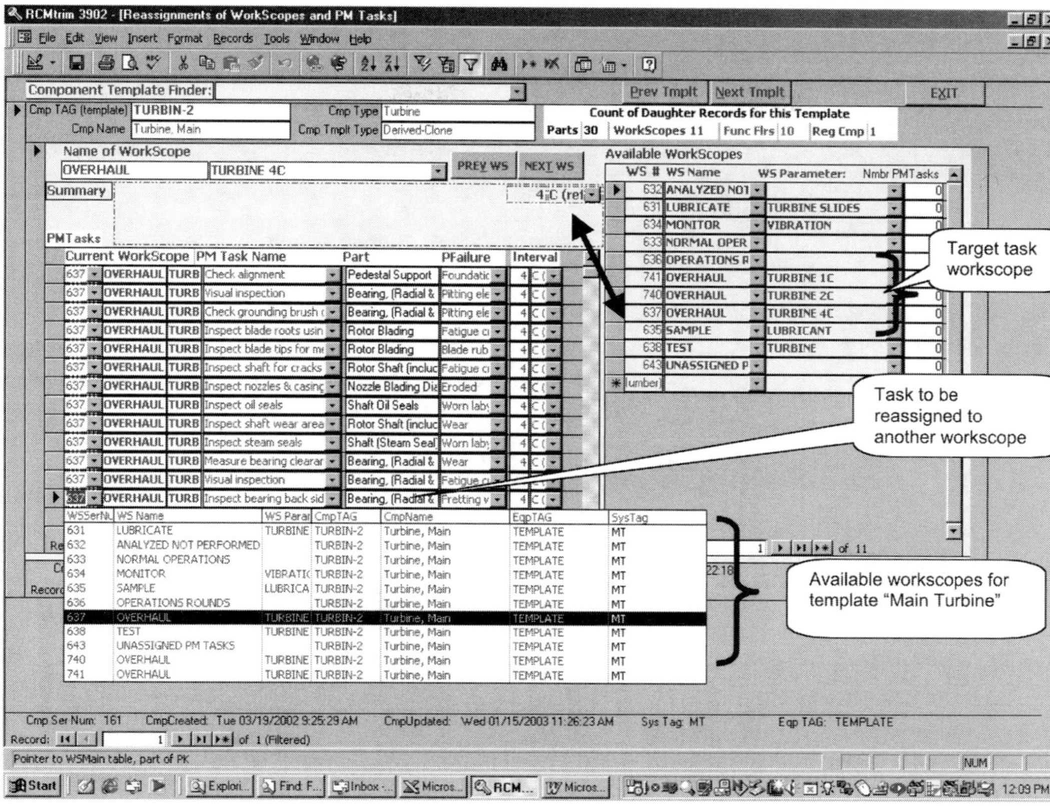

Fig. 6–5 Generic Template Task Reassignment

WO preparation is a discrete job from a work-management perspective. Each WO involves one work site, tagout, trip, work group, and cost accounting charge center. From an efficiency perspective, once scope is defined, it's easy to plan a job as a defined set of activities. From the PM perspective, each workscope PM task performance is well defined, can be estimated with man-hours based on time accounting principles, and can receive a duration for overall running time (*see* Fig. 6–6). Duration is necessary to plan outages involving hundred or even thousands of WOs!

Trip to and from the site, tools/parts pick-up, and clean-up are part of the total work order overhead. WO overhead should be charged to the job. Workscopes define job overheads and allocate those to the job. They also enable efficient task blocking. By assigning overall work order trip time, individual task bias is avoided. Coordinating many PM tasks into one trip encourages an efficient use of worker time.

Trip time

Trip time is time spent going to and from the work site, including breaks, tool crib trips, and parts trips. Overall trip time contributes a major part of overall WO work time. By assigning trip time to a work order, trip time is distributed over the total job,

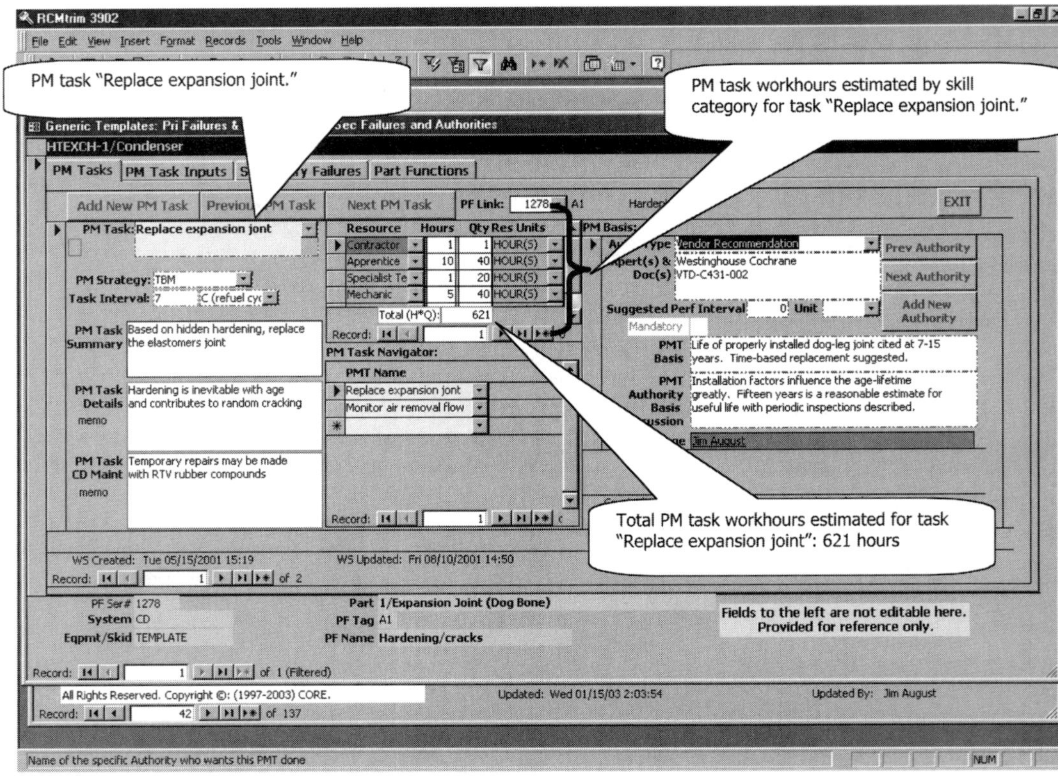

Fig. 6–6 WO Task Craft Time Accounting

reducing the cost and time allocated to each task. As work efficiency increases, trip time drops. By grouping tasks into workscopes, the time per task is reduced. For large jobs such as machine overhauls, the time getting tools and parts, and taking care of personal needs can be more than 50% of the total. Reducing these contributions is desirable to maintain overall job cost low (*see* Figs. 6–7 and 6–8).

Labor values

Labor performs maintenance. It spends only part of its time *turning wrenches*. Benchmark estimates suggest about 60% of maintenance time is spent working in world-class maintenance organizations. Typical values are closer to 40%. Maintenance wrench time is lost to travel, tools, parts, and other lost time "runs." Because even modest improvements in maintenance lost time greatly improves wrench time on the job, blocking offers great work organization value. WO overheads can be acknowledged and dealt with. Anything that better organizes work around WO trips improves overall maintenance productivity. Measuring the contributors to overhead and knowing what their contribution to overall job cost helps to manage their impact.

In a highly efficient environment, more scheduled work is possible because total delivery cost is lower. The more efficient the organization, the more it can afford PM, and the less reactive it needs to be. Thus, cost benefits of efficient work performance accumulate and compound.

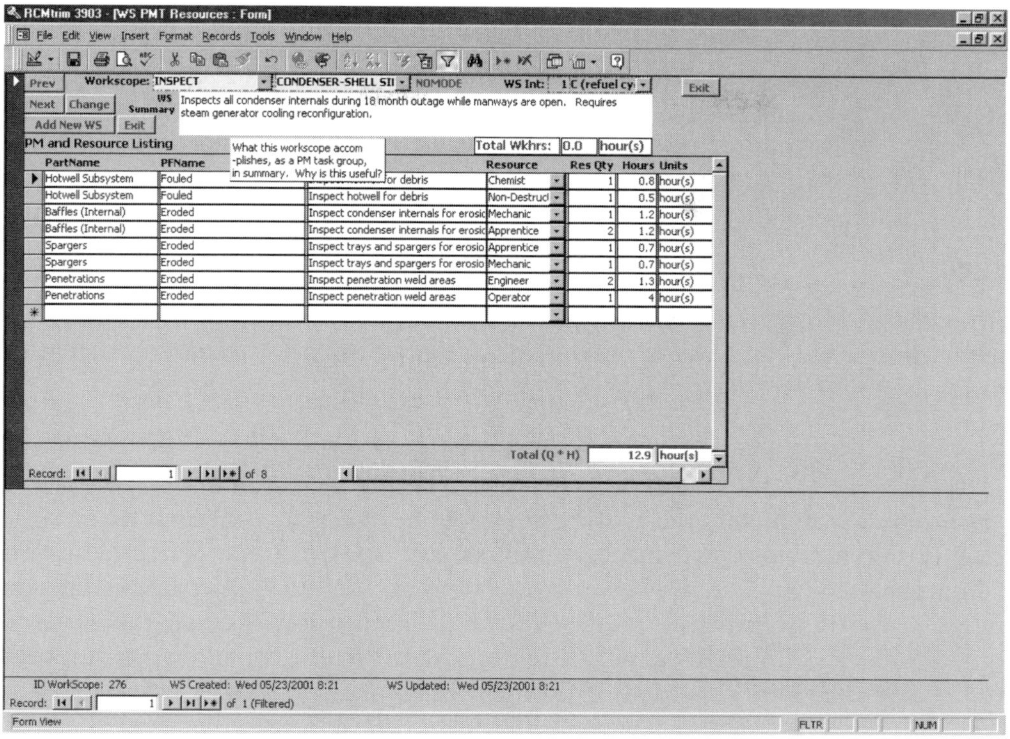

Fig. 6–7 Labor Hours Breakdown by Task

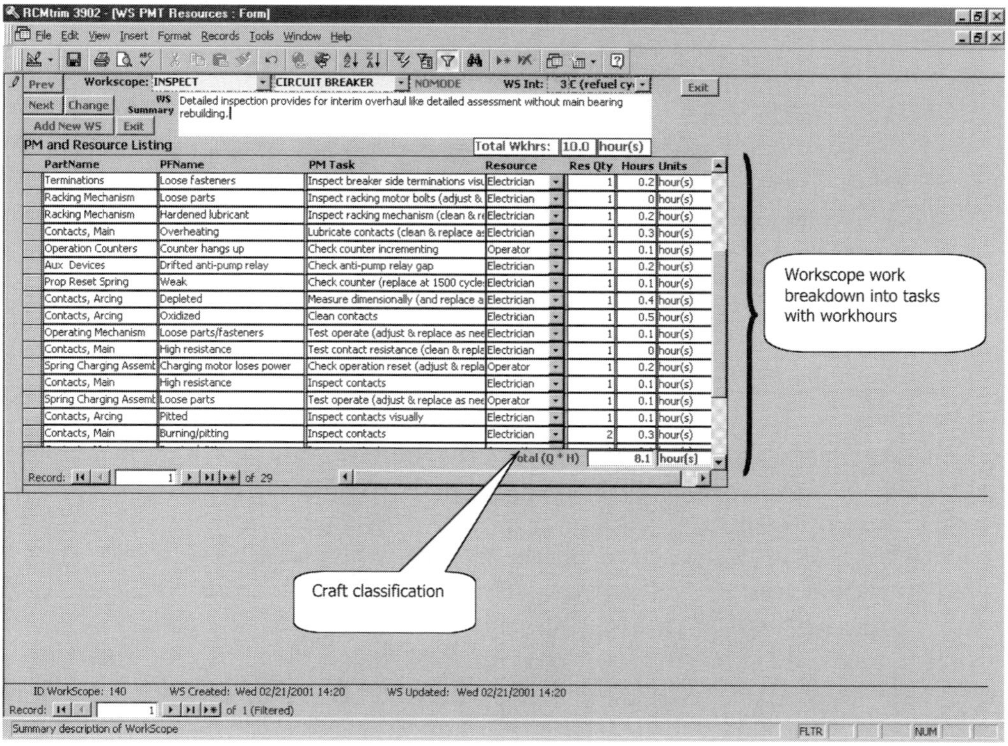

Fig. 6–8 Task Labor Roll-up to WO Workscope

Tools

Tools are a resource, like labor. Special tools are a special resource. Tools that are not generally available should be separately identified and tracked to ensure their availability.

Specialists

Specialists, or experts, are another resource. When scheduled maintenance requires experts, these people can be allocated to the work order like other resources (*see* Fig. 6–9).

Differences in generic and applied template workscopes

For generic templates, the workscope provides a rough container. It provides a starting point for work organization around anticipated work performance tasks. Until a template is applied, however, the scope is not defined. Once applied tasks are selected, scope must generally be re-evaluated based upon selected tasks and new assigned intervals. Until the actual system installation and context becomes known, the correct workscopes simply aren't known. Final organization of workscopes must

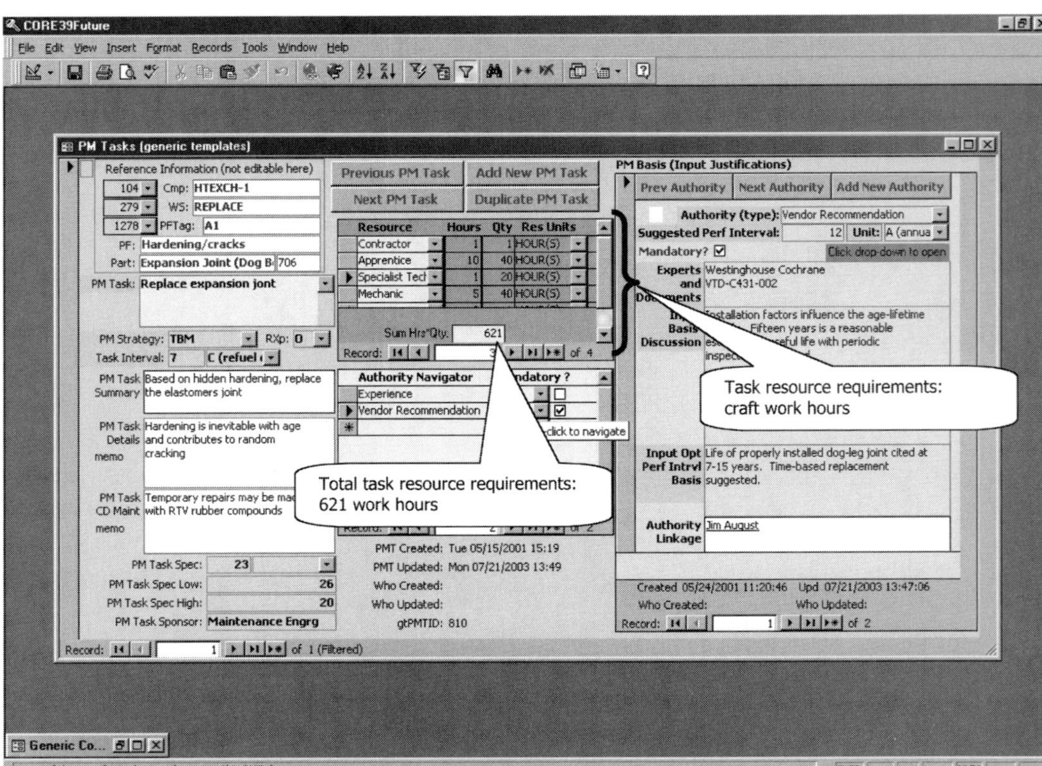

Fig. 6–9 WO Task Specialist Time Estimates

occur on the applied template. Here, tasks can be regrouped, intervals for the grouped tasks adjusted, and the consequent scope of work loaded into the upload tables for CMMS/EAMS use.

Because the individual tasks from the generic template (1) depend on the application context of

- selected tasks
- task intervals
- tagout boundaries
- work availability windows

and (2) require re-blocking once tasks themselves are selected, considerable tuning may be needed to yield a workable block of WO tasks. Conditions also change, as well as the need to re-plan work. Thus flexible techniques to adjust work are needed.

7

Barriers to Practicing RCM

Avoiding PMO Traps: the Dominant Failure Mode

Traditional RCM analysis focuses on equipment reliability from the system down. New system analysis ignores pre-existing work (including current PM program tasks) and their basis. Rigorous, traditional RCM presupposes that existing work biases analysis. Streamlined RCM processes do not accept this assumption. Foundation-up analysis development constrains analysis performance speed *(see* Fig. 7–1).

Failure data used in analysis may be inferred without fieldwork. Organizations tend to justify ("grandfather") their current scheduled maintenance program, although controls can counter this temptation. Managing bias takes planning. Legacy tasks should not be blindly accepted; otherwise, the there is no inherent reason not to consider existing tasks in new RCM program development.

Practically, analysts can conduct their analysis so that hiding legacy tasks origin is virtually impossible. PM tasks must meet and pass applicability and effectiveness (A/E) tests. There can be no PM program task grandfathering in an RCM-based program. Every task must stand on its own failure-prevention merit.

Dominant failure modes (DFM) emerge in plant operations as the failures that matter. They drive reliability and cost. Finding dominant failure modes is an essential part of RCM analysis. Only actual, substantiated failures and their modes should be considered as DFM candidates. Statistics should be used as one tool to identify DFM, but interviews and benchmark comparisons are also required. Engineering design, experience, and an open mind help define DFM for safety, production, and operating costs.

Successful DFM identification takes failure-engineering experience, equipment knowledge, research, inference, intuition, and the ability to work with incomplete information. Not everyone has the patience or skill to perform this work. It requires an

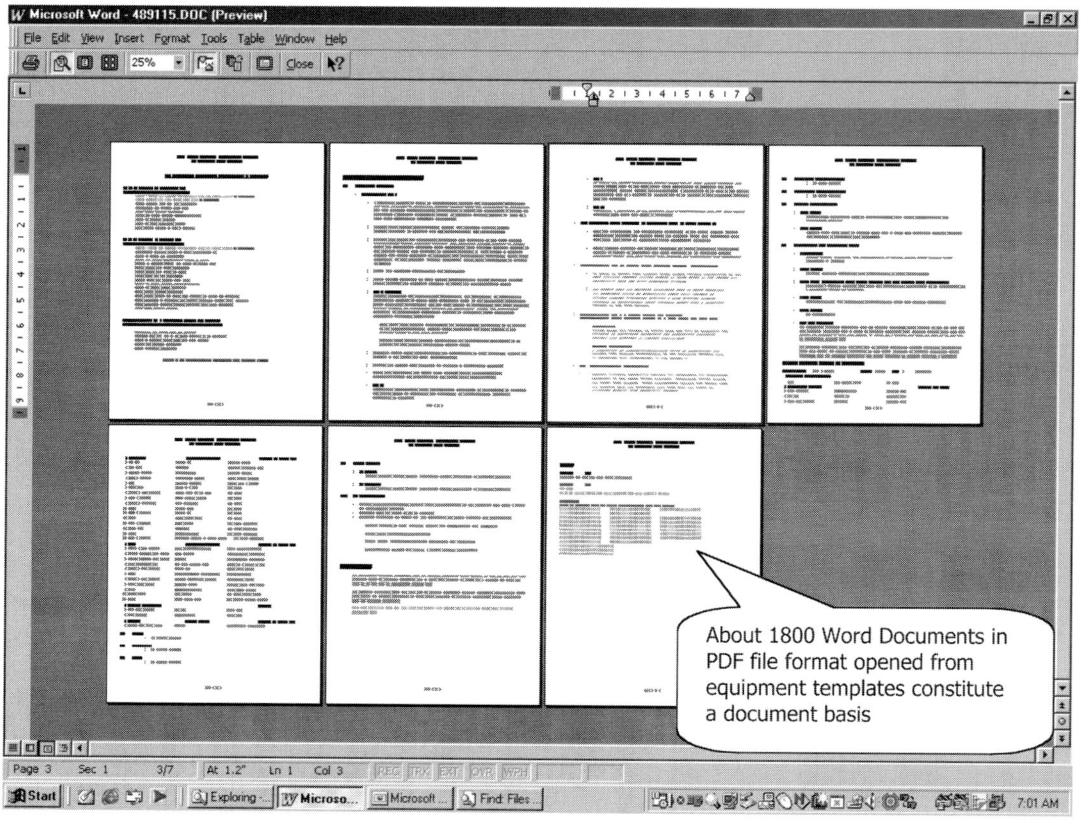

Fig. 7–1 PM Text Basis Documents

engineering background, yet not all engineers have the reliability perspective. Because of cost, work performed should be extended considerably; it's easy to specify all the possible tasks the vender developed over their product development cycle. It's harder to select only those tasks that add value because they are necessary in one equipment context. Deciding what parts and expressed failure modes complement the equipment selection—the dominant failure modes that matter in the equipment under field conditions requires judgment and authority.

Suppliers know their equipment's common failure modes and provide specific inspection points and tasks in their O&M manuals to address these. They do an exceptional job identifying safe life limits, which with codes provide a solid task selection foundation. PM task selection should rigorously define failures and their actionable, effective PM tasks. Careful work group reviews fill voids where failure data is incomplete, but this type of information gathering should be treated with caution.

Although PM task selection should be based on group consensus, it must also pass basic RCM applicability, actionability. *Applicable* means that it addresses a legitimate DFM for the equipment. *Effective* relates to cost. *Actionability* means it can be performed unambiguously in the plant environment. For example, the setpoint limits have been expressly identified, or the part to be replaced is unambiguously identified by a drawing. This criteria removes discretion from the task. Its performance becomes specified and certain. Re-evaluating an existing program without scrutiny or restating existing PM activity with no critical review is pointless. Surely some maintenance optimization efforts have followed this path!

PM engineer/developers especially get trapped enumerating failure modes. Reviewing new systems, engineers and analysts delve into complex inner equipment workings, classifying equipment failures, and developing and presenting findings to plant review teams. Equipment criticality risk justifies and provides the technical basis for scheduled maintenance, applying generic templates. In that context, *apply* means to perform the following set of activities:

- select an appropriate generic template
- pick expressed dominant failure modes—and their associated tasks
- adjust selected tasks and intervals to service conditions

Part-part failure-PM task (P-PF-PMT) reflects plant-installed equipment's actual service, environment, and risk. Generic templates include many tasks, many of which aren't applicable to specific plant equipment. Template applications select generic template tasks relevant to real, installed equipment based on risk and expressed DFMs. Applications adjust task performance intervals appropriately. Theoretically, applied templates could have every generic template P-PF-PMT dominant failure mode associated record string listed, addressing expressed dominant failure modes.

Statistically and practically, installed components don't have every problem ever found, nor do they use all possible generic template tasks. Indiscriminate task application reflects a PM optimization (PMO) philosophy. PMO applies all tasks at the most conservative intervals because risk and context customization—RCM features—are not used (*see* Fig. 7–2).

PM tasks, applied without regard to service, risk, or environment are PMO. PMO bases maintenance plan development on components without considering risk. PMO was developed as a nuclear power PM strategy alternative in the 1980s and has widespread use in the nuclear power generation industry today. But PMO is not RCM! PMO does not

- focus on DFM
- select tasks by failure risk
- consider PM tasks' risk based upon function affected
- consider specific equipment context
- consider secondary failure probability and consequences

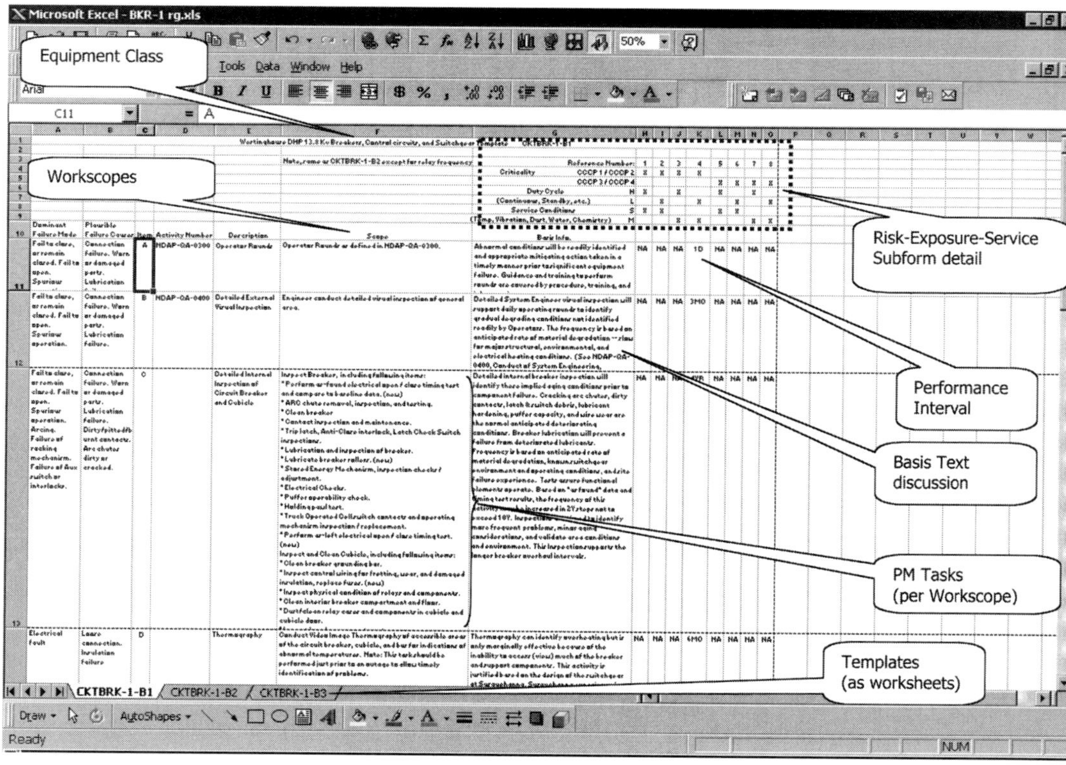

Fig. 7–2 PMO Spreadsheets

PMO results in bulky, applied-template work orders that perform too many tasks. Unnecessary tasks extend field workscope unnecessarily with low- or even negative-value work. Negative-value work introduces infant mortality failures that decrease reliability. A bulky process that focuses on developing and selecting all potential work tasks without specific risk consideration becomes an end in itself. PMO should be suspected when PM templates exhibit lengthy task lists crossing many component types differing in service, risk exposure, and environmental context classes. PMO justifications are text-based, lacking specificity to equipment failure context. Techniques to use to avoid the PMO trap include use of the following:

- rigorous failure statistics preparation
- craft reviews to validate failure applicability
- careful PM task review for clear *actionable* tasks
- critical expert PM task applicability reviews on applied templates
- system-function based analysis
- considerate use of a carefully planned database

Template task application should consider each failure mode by actual failure characteristic and adjust task intervals based on degradation experience. It's tempting to carry tasks forward at comfortable, conservative intervals, e.g., mandatory refueling cycle (18-month) or turbine/boiler inspection intervals (12-month), although outage minimum intervals are the lowest common denominators for long-term work. Actuarial failure data rarely suggest failure controls requiring such short intervals. These default work periods fill routine maintenance work orders and tasks with no firm engineering basis. Workscope blocking is the sole, rational justification for performing the work at the 12/18-month interval. Electronic instrument drift-rates, part wear, or failure history establishes appropriate task performance intervals. Establishing PM intervals requires fieldwork, including craft interviews, physical part inspections, engineering knowledge, and an understanding of equipment use and degradation (P-PF) symptoms. Changing intervals and setting practices takes a bulldog personality to tackle existing habit! It's easy not to challenge existing, tightly conservative intervals lacking justification basis that exists—inappropriately—on so much equipment. Differentiating risk by criticality criteria (SOC) helps the analysts assign tasks appropriately. A failure identified as C, for example, opens the option to extend the interval literally to no scheduled maintenance (NSM). This enables analysts to more appropriately address cost.

Some streamlined RCM advocates emphasize alternatives to statistical failure data collection. Mark Twain said, "There are lies, damn lies, and statistics," but failure data statistics—be they damn lies or worse—are the best this world has to offer. Where statistics are available, they should take precedence until better statistics, operating examples, or other cases are developed. There can be no voodoo in RCM; scientific approaches should always outrank off-the-cuff, seat-of-the-pants methodology. At RCM's origin, Nowlan and Heap demonstrated by means of statistics the futility of *hit-and-miss* time-based PM selection approaches that were contrary to the public interest. Though pursued in response to public safety, and not obvious at the time, regulated PM work promised merit but lacked engineering basis. The corresponding intrusive time-based activity caused more failures to apparently adequate equipment than they corrected based upon Nowlan and Heap's trail-blazing statistical assessment. Although the input of the craft people is crucial, policy should weigh all inputs with sound engineering practice. Scientific facts must prevail over opinions. Statistically sound tasks reflect good science and engineering.

Some RCM analysis processes assign more credibility to group discussion and opinion than statistics. This contradicts a painful practical lesson learned several times over in direct plant support: Groups are highly biased towards recent events, and memory is inexact. In legal practice, statutes of limitation reflect the lack of memory accuracy. The courts cite testimony versus hearsay and argue for the need to weigh all evidence in a balanced forum before so much time lapses as to make the reconstruction of an event fruitless. Statistics, therefore, anchor opinions with reality. People are tempted to create their own reality, but statistics provide a foundation that intrinsically filters hearsay. Plants are verbal cultures that create their unique plant myths. Statistics can keep companies from making inappropriate operating decisions based on hearsay alone when far more reliable methods can (and should) be used.

Incremental improvement

Traditional nuclear plant practice tunes existing processes with minor midcourse corrections. Workload trimming—selectively focused on streamlining PM in the context of overall maintenance process improvement—is an attractive yet iterative improvement. Iterative improvements produce modest results. Small changes are more readily blocked and opposed organizationally at multiple worker levels. Non-radical changes require no committed organizational action. Lack of enthusiasm easily bogs down daily implementation and use.

Vested parties must buy into streamlining. Successful project outcome must supersede individual and even department sub-optimization objectives.

In contrast with modest commitment, step–changes—especially those with cultural change demands—require greater commitment and risk more. Threatened parties are more visible and active. Benefits and overall outcome risk are increased with total process change. Management must get behind structural process change from the top. All managers have to believe in the change. From this belief, opportunity for substantial performance improvement arises.

Analysis performance

Performing traditional power generation RCM analysis requires time to perform analysis. Industry thumb rules suggest 200 hours are required per system for contracted RCM. Unfortunately, such estimates are nearly useless; average systems are estimated at about 2,000 equipment tags in nuclear plants with 200 in fossil plants; but numerical components vary from 200 to 10,000+ per system at nuclear units alone! Electrical distribution systems run around 5,000, and plant controls number even more. The level to which components are tagged varies within plants.

System definitions themselves are arbitrary and vary tremendously within companies and even within the same plant. Identical plants may be broken into 50 systems at one location and 150 systems at the next. So *average* has little meaning. Factors to consider when determining analysis time include:

- level of detail
- explicit basis requirement
- statistical failure analysis
- total component numbers to review
- total critical components requiring detailed analysis
- suitability of standard generic templates
- onsite expert review
- analytical perfection and customization
- tool availability and ease of use

Where perfection isn't required, analysis *can* be completed in 200 hours per system, using standard component templates, standard systems, expert reviews, failure validation (instead of statistical WO trouble reports, operational events, and failure analysis), and limited basis for input-tailoring the number drops. Where exact analysis *is* needed, the sky's the limit! For new projects, 1,000 hours per system isn't unusual to train fresh plant staff in the performance of basic RCM process analysis in production. Costs should balance with consideration of who does analysis and station performance objectives. Contractors can perform analysis faster at lower overall cost, but technology transfer to the users who must make the product work will be limited. Learning RCM is intensive.

Where the purpose of the analysis is to transfer technology and learn process methods, lengthy studies of a few systems should be anticipated and sought. The real objective on any project must be achieving a sustainable learning curve and performance improvement by some increment (~10%) with each succeeding system analyzed.

Cost perceptions and consequences

So, if RCM analysis is reported to require 100 man-hours per system—and these numbers may be optimistic—some efforts are not included in these numbers. One hundred man-hours per system translate to $10,000 per system at nominal rates. Plant analysis costs run around $500,000 per 50-system plant. These costs are intimidating to those working on limited budgets. Further, they don't reflect the full cost of RCM, which must include implementation.

For more widespread use, RCM analysis costs must drop and benefits rise. This requires a cultural shift to critically evaluating the basis for work in many plants. North American maintenance cultures democratically share rights to initiate work at all levels regardless of technical merit. The shift away from this position is profound and affects the way maintenance people view their jobs. Maintenance in this new paradigm orients to production, becoming more of a production enabler and less of an end in itself. Understanding maintenance performed becomes as important as performing the work. For most traditional maintenance workers, this shift in philosophy is profound.

How far maintenance has come from the simple task of picking up work orders each morning after a cup of coffee in the shop!

Legacy programs

A legacy program that maintains reasonable performance using traditional methods is an attractive alternative to RCM. Why risk upsetting the apple cart with a new approach when the tried and true works?

This logic challenges increasing applications of RCM-based ideas. A good PM optimization-based program is the *de facto* benchmark standard today. Best-in-class plants meet this operational standard with their CMMS-based PM programs. Legacy programs lack

- systematic basis
- age exploration
- risk assessment
- failure-based PM tasking
- workscope replanning flexibility
- task-level PM cost accounting
- PM reliability value measurement
- PM reliability engineering interface

If plant performance is satisfactory, there's little purpose exploring this issue. However there are many indications that opportunities for improvement abound in many industries. Maintenance remains highly crafted. Maintenance task bases aren't clearly understood, and the contributors to maintenance cost aren't known.

While nuclear plants struggle with the need to maintain a basis for regulatory reasons, the challenge in other plants is to maintain enough bases cost-effectively to sustain maintenance programs that continually reduce costs. Legacy programs founded on PM optimization (or even earlier programs) lack that capability. RCM, in contrast, implemented with the latest relational database design, offers the best opportunity to systematically tackle maintenance costs while improving reliability.

The utility industry is conservative. As the first bout of deregulation evolves, the challenge for many is convincing regulators that they remain on the cutting edge maintaining the very best operating and maintenance programs. RCM will continue to *be* that cutting edge approach to maintenance for the foreseeable future. Its replacement—real-time condition monitoring—looks very promising but hinges on the successful task selection of RCM itself! There is little likelihood that companies will transition easily to the next level without first finding an interim stepping-stone. RCM provides that step. For many, that interim step is too hard to visualize even now.

To speed the cost of legacy program conversions, several techniques build shells with the legacy material and then expand outward by filling in the missing reliability gaps to greatly improve process speed. Existing basis materials can be added to explicit basis fields quickly, preserving detail legacy analysis while preparing the facility to move to a higher analysis level. Cost of maintenance program maintenance is the fundamental justification most companies use for new processes or software. Although the learning curve for the traditional organization is steep, when converted, software-based RCM-grounded PM programs are easy to maintain and provide a permanent tool for reliability, cost, and production improvement. The engineering and maintenance productivity they provide recoups their cost in less than two years!

Excluded middle SOC risks

The trend in maintenance programs is towards two tiers of risk—safety and cost. Practically, elevating safety to identify user *wants* as a stepchild for funding has been in practice for years in many plants. Workers simply phrase the basis questions so the answers can be easily provided in safety format—whether or not there is a direct safety basis. *Gatronix* public address phones nearby, new convenience stairs, convenience access ladders, more handrails and even electrical receptacles that reduce costs are elevated to a safety basis when their proponents really want them in the plant budget. Workers know they fair better when their requests are made with safety spin.

Many years spent with these ineffective work categorization systems have left a bifurcated risk world. First of all, safety is sacrosanct. The direct safety condition of aerospace RCM vanished (if indeed it ever existed), and many layers of safety maintain excessive redundancy levels that raise costs. The risk of too much is worse than too little. Too many redundancy levels are more difficult to understand, test, and maintain. Crews become conditioned to redundancy levels. They loose track or ignore redundant layers until those layers fail over time, are effectively lost, and an operating event occurs *with multiple failed layers*!

Layer depth increases complexity. Complexity inevitably leads to failure randomness in the absence of a conscious, coherent strategy to maintain them. Without concise, comprehensive plans and strategies, there is simply no systematic way to remove random failure from operations.

Characteristics of RCM PM changes

Most existing PM program changes come as minor tasks formally added to the existing program. Task deletions or interval extensions require responsible owners to grasp the underlying ideas and take fundamental ownership. Developing confidence limits and skills requires time.

Setting intervals requires a combination of operating confidence and risk awareness skill that individual engineers must develop. The risk of canceling value-adding PM work is very slight. With risk-critical exposure classes identified explicitly at all levels, the risk of safety or operations errors diminishes. Where program task cancellation or extension occurs, the infamous final sanity check asks, "Do we really want to do this?" At this point, if there is any doubt, experts can examine changes more closely.

Quality considerations

Several checks validate RCM analysis and improve quality. The most important include the craft and responsible engineer reviews. Those with hands-on experience with the equipment readily identify errors. Other errors introduce work they can't intuitively see as valuable. Both checks validate steps before implementing the final program. Errors of commission are more likely to develop in regulated PM programs, where "more" has an implied meaning context of "better." Additions to these programs are easier to implement than consolidations.

Review

Maintenance performers, engineers, and systems teams are expert reviewers. Final PM task customization captures generic template tasks, making template application installation-specific. Detailed knowledge and experience gleaned over time quickly become available to the entire plant. Process reviews reduce the likelihood of error. Most task changes are new task additions. Scrutinizing task removal several times provides assurance against over-zealous cutting.

Further, the sanity check independently double-checks implementation recommendations before an actual change performance—including task removals—occurs. When performing both checks, results can be compared. For identical results, the faster process is superior. In these ways, streamlined RCM plays well to its legacy antecedent.

Process Considerations

8

Upload

Uploading the RCM-based scheduled maintenance program from the analytical engine to the CMMS/EAMS completes the analytical work. Uploading depends on the type of CMMS/EAMS where the final scheduled maintenance plan will reside and on the format used to develop the new PM program. Text documents like those produced with Microsoft Word have been in use for a least a decade to develop and maintain new scheduled maintenance plans. They are easy to develop, but uploading the results is limited to providing reference documents to be re-entered into the CMMS/EAMS scheduled maintenance table text fields. At best, Word documents with relevant PM plans can be opened and information cut and pasted into the scheduled maintenance tables.

CMMS/EAMS systems in commercial use today use relational database engines like MS SQL Server and Oracle 9i. They transfer PM table data files into their scheduled maintenance tables with straightforward queries that can import data from spreadsheets like Excel or databases like Access using an *object database compliant* (ODBC) process like that provided by Access. The key to successful uploading is developing the output products so that uploaded files are compatible with the table formats of the CMMS/EAMS. Every CMMS/EAMS has its own location and use design for the location and use of scheduled maintenance tables. Some use PM models to initiate new maintenance work orders as they are scheduled in the PM subroutine. The model is essentially the content of the work order. Scheduling functionality is separate and triggers the reissue of the model creating another work order based upon a time event in the CMMS/EAMS software application. This is like having a master copy of a file document that gets copied, annotated with control information (number, time and date, assignee, time...), and issued for work, which is how work proceeded prior to the advent of CMMS/EAMS systems in the 1980s.

How CMMS/EAMS systems create and track new work orders depends on the specific application. Practically, this means a special interface is required to match external scheduled maintenance development packages with any CMMS/EAMS. Even when work is done in MS Excel spreadsheet format, output table(s) must be constructed with the final table locations in mind. Data mapping of the results between the two application locations is a preliminary step (*see* Fig. 8–1).

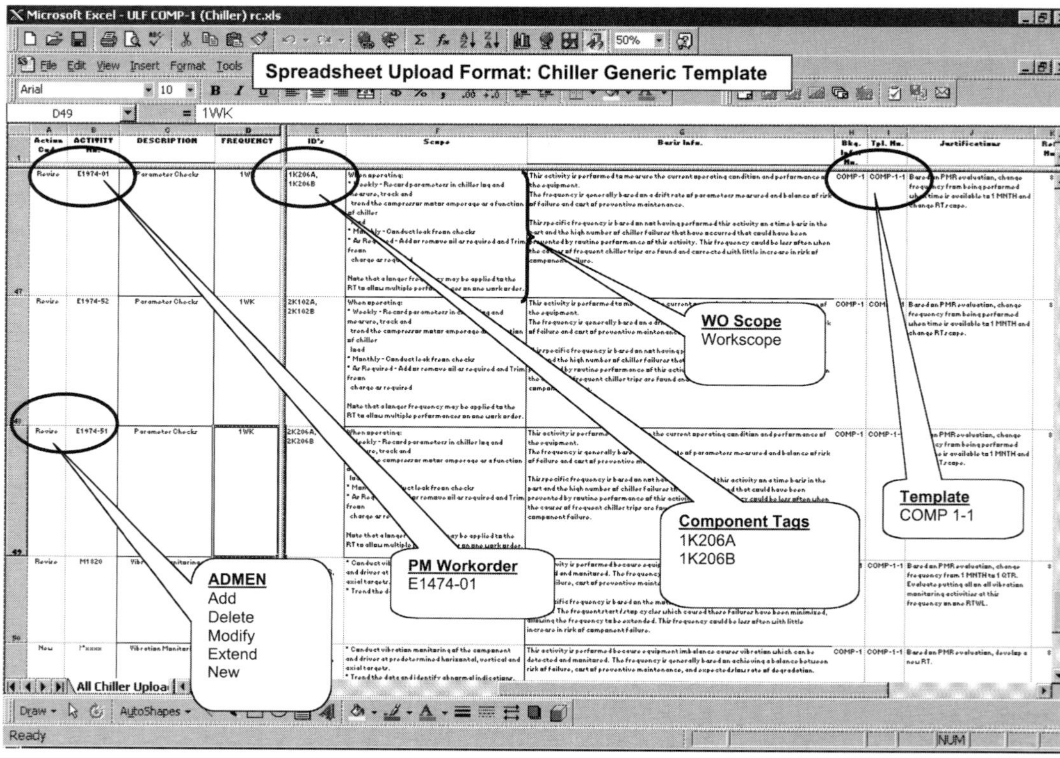

Fig. 8–1 PM CMMS Upload File Spreadsheet

Uploading results also calls for careful consideration of how the new program will overlay the old. If you individually compare old WOs to new WOs, there are five potential results.

- New WOs can be added. (Add)
- Old WOs can be removed as they become obsolete. (Delete)
- Existing WOs can be changed for content or administrative reasons. (Modify)
- Intervals can be adjusted with no content change. (Extend)
- Nothing at all changes (No change)

Summarized, these categories create the acronym ADMEN: Add, Delete, Modify content, Extend (or shorten), No change (including administrative non-content changes) (*see* Fig. 8–2).

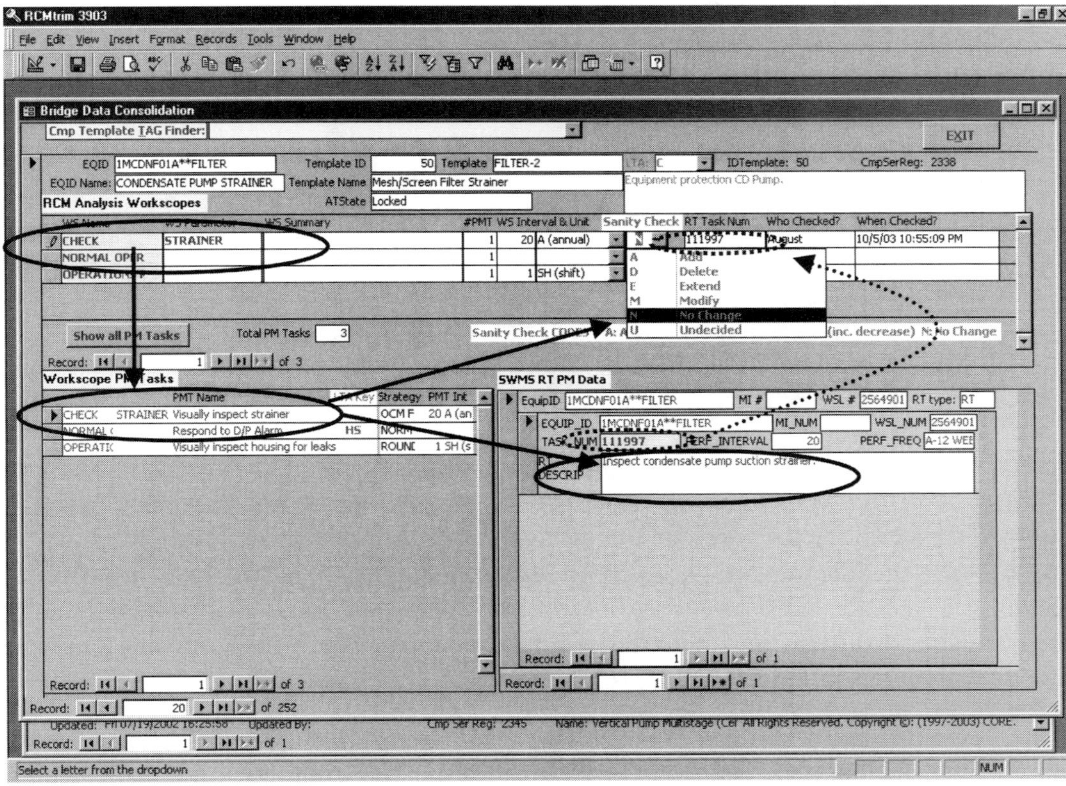

Fig. 8–2 WO Upload Validation "Sanity Check"

For statistical and traceability purposes, it's often desired to do one final check as new activities are placed in the CMMS/EAMS tables. Further, to guide the CMMS/EAMS software in the change modification of PM-WO task content, it's wise to compare each equipment tag's intended scheduled maintenance plan with the revised plan. This creates a list of change codes to allow the CMMS/EAMS middleware to create the new scheduled maintenance tables. Thus, an activity that is coded *D* is deleted and removed to the CMMS/EAMS history file where it is no longer available for use. Conversely, an activity marked *N* (new) is given a primary key for tracking through the CMMS. Those two commands are simple. Modifying category content requires a knowledgeable person (like a planner) to separate the old activities and evaluate them against the new content, task by task.

The task of moving the revised PM program's WOs and their content (tasks, intervals, workscopes, associated equipment, symptoms, limits, etc.) into the CMMS/EAMS is compounded by the design of the Legacy CMMS/EAMS-WO job plan or content fields. Early CMMS/EAMS systems lacked the idea of workscopes in their design; many still don't differentiate a job by tasks. Job breakdown by task is a relatively recent scheduling concept that allows an activity to be specified exactly by task content. For PM purposes, a task is a discrete unit of work related to the management and control of one failure mode. A typical turbine overhaul comprises 20–40 scheduled maintenance PM tasks like blade cleaning, crack inspection, root tip crack inspection, rotor bore inspection, etc., interspersed among a larger WO that controls the disassembly, reassembly, and on-condition tasks (which typically are many) like replacement of failed stage thermocouples.

An objective of RCM is to break down large work like overhauls into their discrete tasks and to relate task-prevented failures to an individual failure-prevention assessment basis on merit. Upon completion of this breakdown-to-tasks comparison (during upload), the comparison against the original work must be completed, if management wants a summary of changes.

Many companies seek assessments for a variety of cost accounting, business planning, and work analysis purposes. By tracking the content of each task strategy (hard-time, on-condition, failure finding, etc.) a historical benchmark breakdown of unit overall scheduled maintenance *completion* can be developed. With each successive batch of PM task WOs uploaded to the CMMS/EAMS, the plant's maintenance program subtly shifts. Over time, these shifts can show overall realignment from hard time to on-condition maintenance.

A general trend in the use of RCM is an increase in *failure finding* and condition monitoring changes that influence operations, cost, and common outcomes. Trends showing availability increase or reductions in maintenance-preventable function failures would be indictors of success for the PM program change.

If data upload into the CMMS/EAMS never occurs, regardless of the analysis effort, no benefit accrues. An upload file provides management with assurance that analysis generates implementation value. Where activities can't be implemented, CMMS/EAMS upload status tables indicate lack of completion. Management can then make appropriate adjustments. Occasionally, for example, tasks are considered for which a company lacks implementation technology or training. These won't get implemented, and their *hold* status forces the question, "Are they worth the money?"

The upload file provides the final CMMS audit trail. In non-regulated environments, an audit trail is simply management's tool to assure job completion. In regulated environments, such as nuclear generation, an upload file provides justification that assures everyone that the maintenance process is managed, meeting standards like ISO 9001, and that scheduled maintenance change implementation is a quality process. All changes are traceable to normal models. Changes have a rich failure justification basis. Every task stands on its own merit, buttressed by engineering interpretations and requirements provided by credible authorities and legal codes (see Fig. 8-2a).

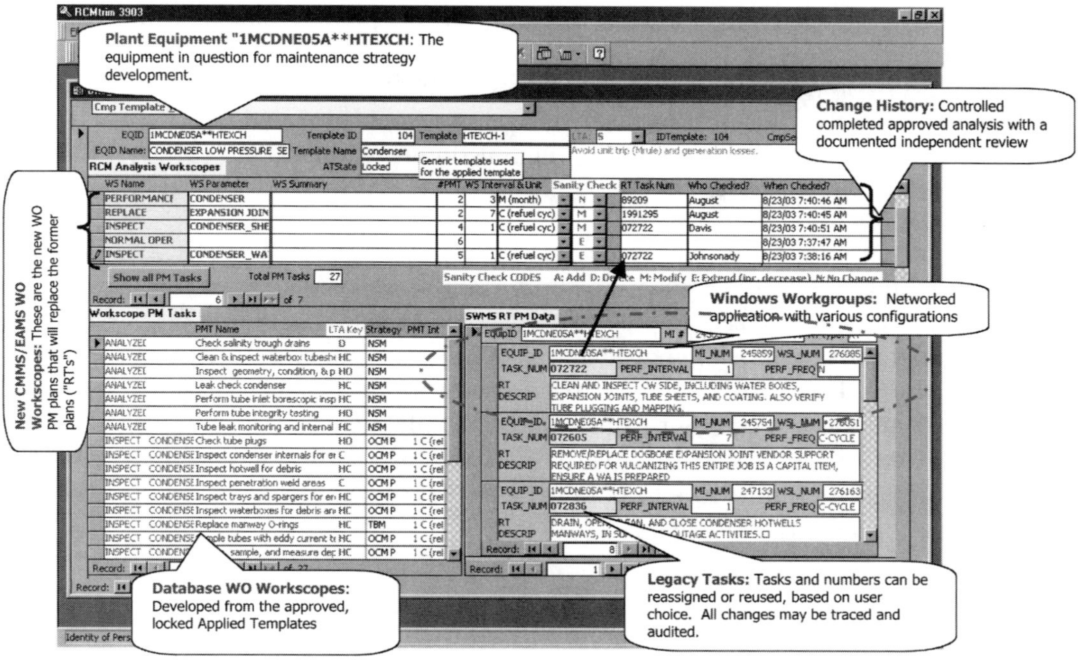

Fig. 8–2a Sanity Check Display

Implementation closes the loop. Downloading equipment tags from the CMMS/EAMS starts the RCM analysis process. The final step is to upload a new set of modified PM tasks to replace and supplement scheduled maintenance WOs. The resulting workscopes can be completed with planning, inside an electronic data environment. At this point the process has come full-circle.

Quality control

Maintaining quality analysis requires standards, group participation, and commitment. Data control ensures quality data management—a plus. Work plan traceability to origins—both technical and regulatory—is an added bonus.

Error rates measure final product output quality. Errors can be measured as a number of internal data inconsistencies, based on the program design. For example, in using one streamlined RCM method discussed, any critical component tag should have an associated normal model and applied template. A simple data inconsistency check of 100,000 component tags at a nuclear plant indicates the statistical error rate.

With measurement in mind, program management can set an error target goal of less than 1% as a general PM project objective. While this comes nowhere close to *six sigma* control rates demanded in manufacturing, for maintenance programs, this is a world-class goal. For perspective, nuclear maintenance programs completed in the

1990s using spreadsheets experienced rates upon completion ranged from 30–40%! Upon program completion, only 60% of prescribed tasks making it into a CMMS/EAMS application were finished without further modification. Not that these were poor programs. They just lacked bulletproof consistency and traceability back to their source document basis without ambiguous interpretations and turns required by an auditor.

Outstanding programs can meet ISO 9000–9003 certification criteria. For scheduled maintenance programs to achieve exacting levels, processes must be clearly defined, interfaces minimized, and performance measures established. While programs cited represent the best available products of their day, they were developed with Lotus 123 or MS Excel-based processes. Experience shows that these processes were limited by what they could provide with the controls available. Modern database products such as MS Access and MS SQL Server trace data back to original changes. Enforced data integrity rules simplify due to data trails leading back to source documents. As a result, process accountability soars.

On the tag count low end, complex chemical, process, and generation facility maintenance systems account for more than 1000 equipment tags and tens of thousands of records. On the high end, nuclear plants with multiple units commonly exceed 500,000 tags! Managing a living project PM process with spreadsheet technology is virtually certain to encounter serious quality control issues. Only access-controlled, user-tracked, managed relational databases can ensure CMMS/EAMS scheduled maintenance database quality.

Consider whether banking (with a simpler scalar data process) could manage 100,000 accounts using spreadsheets! Not to pooh-pooh spreadsheet applications, but the banking example only illustrates the futility of attempting to work with an inadequate tool. In spite of this, most large companies develop and maintain multi-billion dollar facility PM programs with spreadsheets to develop their PM programs. Operating companies procuring services should evaluate all software methods and technology available. Spreadsheet projects have inherent difficulties as well as life limitations concurrent with the project effort. Maintaining spreadsheet basis integrity over the long haul is likely fraught with frustration.

Keeping electronic data in electronic formats without recreating entries is the second quality control issue. Data errors are compounded when data must be re-entered. A general application objective should be to remove, or at least minimize, text entry. Previously entered electronic data should never require re-entry. For RCM development software, this single criterion separates the men from the boys, metaphorically. There are numerous RCM products available, but those that require users to enter their equipment tags aren't serious contenders in industrial maintenance. Using this filter, the discriminating buyer clears the playing field.

Once work is completed, the second opportunity to introduce error is the pass from the RCM/PM developers—the engineers and analysts who perform analysis—to the planners and schedulers (typically in the maintenance department) who input the results into the CMMS/EAMS. Again, to enhance quality, data re-entry must be minimized. Aside from keystroke errors (which are legion), RCM analysis faces a much more insidious challenge: Those entering data in most plants are the historical program owners who interpret meaning at the point of entry. Practically this results in a second gauntlet for RCM analysis, one very likely to make adjustments that takes the program away from its firm engineering foundation. Every major RCM project encounters this roadblock. A bridge software/middleware to facilitate and manage transition upload to the CMMS elegantly addresses the combined need for independent review by fresh plant people, data traceability, and change control.

The bridge facilitates questioning of any specific PM task change going into a PM WO on its technical merit. Because the uploading middleware is traceable to the user, any assessment of or change to the criteria—acronym "ADMEN"—traces back to the point of application.

By means of the middleware, activities like the CMMS/EAMS upload bridge software monitor changes in content, disposition, or other factors. The results are fully auditable and can be checked for arbitrary as well as approved changes. The bridge software interface completes the accountability circle assuring that only approved, controlled PM changes enter the CMMS/EAMS scheduled maintenance tables.

One final quality benefit from local area networks is their ability to remove paper document routing from the program maintenance and development cycle. Users can access, locate, and print reports they need in hardcopy. Where update input and development approvals are needed, these can be provided electronically. The ability to place all relevant information on a network, accessible to users in either browse or (appropriately controlled) update mode, allows data management in a controlled, efficient, friendly way.

The ability to process and develop (and redevelop) maintenance strategy seamlessly in a networked environment supports quality. Achieving this feature is difficult, and the significance is easily lost in practice. However, when achieved the improved user environment represents a substantial process benefit.

Normal models

Identical equipment contexts don't required re-evaluation and reapplication. Rather, they reuse the same work applied template from a previously developed component—the normal model, which is a force multiplier using design symmetry commonly occurring in conservative design. Plant design is repetitious. Normal models accommodate unique designs, as well as operating practice equipment asymmetry, modifications, replacement, or other activity that deviates from perfect symmetry.

A 1980 fossil unit with Foster Wheeler-type MBK mills provides a case study in design control loss. It had evolved by 1994 to have essentially six unique mills with various combinations of air ports, tables, gear boxes, drive motors, dust suppression, and classifier arrangements. One can imagine the operating and maintenance difficulties these posed! Practically, every mill had a different combination of parts reflecting highly crafted designs. Design control had been lost.

Scheduled maintenance accommodation at this plant was extremely difficult. Parts were difficult to locate, and surprises invariably emerged when mills were apart. Design was a mythology based on the last work done. Work orders were superficial documents. Buyers bought parts from vendors for many reasons other than the primary ones—meeting specifications and providing value. In reality, this plant had six normal models for coal mills where one or at most two should have been adequate. One normal model would have reflected perfect symmetry of design and use. In fact, the sixth mill was slightly different in tuning and use. It provided a ready standby for the other mills, and the plant was capable of operating on five or six mills based on primary airflow bias.

The normal model provides a reusable, exact model of a maintenance template where symmetry of usage, service, and design are present. Slightly modified—and where empirical evidence indicates no asymmetry of these three attributes is present—it should be presumed that service is symmetric. The normal model can extend to items that are not exactly the same, except for dominant failure modes and key failing parts behaviors. The normal model is hard to grasp without design symmetry appreciation and operational use.

Maintenance people seldom learn the operational side of equipment. Operations clears their equipment, and that's as far as their training goes. The best instrument and control (I&C) technicians, on the other hand, are former operators with loads of design appreciation. They learn operations working on plant controls; failure to understand the overall control scheme often means an inadvertent plant trip! While it's not necessary to teach each maintenance craft about operating risk, there is benefit in helping them understand how standby or out of service standby equipment contributes to plant trip operating risk.

One specific case involved a three-pump arrangement where two pumps contributed 100% of critical flow. The third pump is, if anything, half a spare. Maintenance treated it as a full standby, trying to perform work online, with repetitive trips that operations was unable to resolve due to departmentalization. Operations and maintenance could not communicate nor act as a team to develop a common strategy to manage the equipment for the plant's availability benefit. While these organizational missteps provoke thought, the daily reality is that if unresolved, production suffers. Lack of resolution stems from problems ignorance, lack of root cause understanding, an unawareness of fundamental design issues, and the lack of resolve to realize the maximum operational benefits that design can offer.

The corollary is that savvy operating companies know enough about plant design—equipment redundancies, maintenance layout space, quality equipment requirements, train redundancy, procurement practices, and continuous improvement—that they don't make these legacy coordination mistakes. Their performance is world class, and that's what separates the operating companies from mere owners. More than ever, companies are making decisions to become operating companies or asset managers with operations run by professional operating companies. Being halfway-committed results in lukewarm results.

A practical problem implementing normal models is that most maintenance departments have worked with PM optimization. Achieving scheduled maintenance improvements requires more participation from engineering than historically provided. Engineering costs are prohibitive, so there must be a payback. Few engineers have the operational maintenance background of a plant engineer. Company-wide implementation tools help make design-to-maintenance reliability transition easier and improve plant engineers' productivity. These implementation tools assure faithful design representation in the scheduled maintenance plan's intended requirements.

Cost

Costs present a dilemma to engineering groups in large organizations. Engineers exhibit passing interest in costs; they like to do design! Plant managers, lead engineers, or maintenance managers are charged with figuring out the complexities and nuances of plant cost. Cost development appreciation occurs later in technical careers than is useful for many companies. Some engineering groups reject all responsibility and interest in cost management. Unfortunately, costs and engineering are closely related: If some engineers had a better grasp of cost-benefit relationships, some activities no doubt would have not been performed, while others left undone would have been completed.

Cost engineers are unwelcome in industries that lack cost-management discipline. Lawmakers identify idealized rules about emissions and safety checks, but the companies implement these rules. Where costs and safety clash, everyone hopes for the best.

Thumb rules—back-of-the-envelope cost consciousness and skills—are highly sought in competitive industries. Utility and government engineering departments depend on thumb rules less, or they may even be made irrelevant. Cost discussions question the nuclear safety analysis focus. Of course, rising nuclear operating and construction costs contributed as much as TMI to the dearth of new nuclear orders in North America over the past 25 years.

Catastrophic events and major problems prompt plants to rework equipment, facilities, or budgets. In the face of events, unbudgeted money often is made available even though unbudgeted. Thus, $2 million boiler panel replacements, $5 million condenser re-tubing, $10 million generator refurbishment, and $20 million control rod drive refurbishment are paid out of general corporate funds and never reflected in plant budgets as formal expenses because of the unusual, one-time event status and funding surrounding these expenditures. Catastrophic failure event costs, then, never fold back into plant operating plans as real operating costs.

Expenses required to achieve reliable base-load operations following catastrophic events never complete the plant's total cost or income budget picture. Some managers live in a production and operating-cost fantasy world. Without knowing true operating costs, real benefits of improved scheduled maintenance practices are difficult to assess.

Familiar benchmark cases provide references. High-profile events like generator winding failures create lasting impressions. These events convey corporate values and messages about safety and can appropriately convey cost. *Foobars*—events overlooked that led to errors—usually involve secondary failures. Programs that should have prevented the primary failure, like instrument alarms or fan trip, but were often part of the root-cause picture that created the event were lost. Industry failure data helps judge basic PM task failures on non-critical equipment.

For maintenance, two types of PM analysis of cost problems follow the 80/20 rule: the important few and the valuable many.

Important Few

Two occurrences drive overall PM program value: forced outage events that cause lost income, and events, like high-cost equipment failures. Equipment failures that cause plant outages and outage-related costs quickly spiral into large, direct expenses based on the resulting production losses.

Equipment failures that only cost repairs (with no outage basis) rarely exceed more than a few million dollars for even the worst equipment losses. Compressor or pump failures, waste concentrator failures, partial cooling water tower collapses—these don't force outages, but they are expensive. A thumb rule for evaluating substantial rework repair costs is $50,000. If the repair cost exceeds $50,000, they are significant; if less, they aren't. With this threshold, examining some benchmark occurrences with great value and payback benefit serves to anchor cost-based PM decisions.

Sootblowing air compressor (SBAC) filters

The intake filter replacement set on a 15,000 acfm SBAC air compressor runs around $1,000 for materials (filters), another $1000 for labor. (ACFM means *atmospheric cubic feet per minute* with volumetric flow calculated at atmospheric conditions.) Not replacing intake air filters in a timely way shortens finishing-stage compressor wheel life due to foil erosion. (Plugged filters tear out.) This reduces the potential life of the rebuilt compressor by 50–75% where airborne dust is present, like the western United States. The value of the *realized* potential compressor life is around 50% the cost of overhaul. Overhauls cost $500,000 every four years—the practical, realizable life based upon perfect filter replacement and other operational considerations (*see* Table 8–1).

Table 8–1 Sootblowing Air Compressor (SBAC) filters

Action	Costs ($)	Avoided Cost ($)	Benefit ($)	Cost/Benefit (benefit/cost ratio)
Replace filter set	2,000x8	250,000		234/16
	16,000		234,000	
				14.5

What most people suspect intuitively is that there is substantial cost benefit in avoiding premature overhaul by performing appropriate manufacturer-suggested PM. PM completion can be measured easily and adjusted to plant operating requirements. Several additional subtle points are the following:

- The threshold where PM tasks are mandatory based on cost is the 5–10 benefit-to-cost ratio range. Equipment crashes are not just painful out-of-pocket expenses, they drain many resources. They drag down other work down causing programs to fail, including scheduled maintenance programs.
- Consequential effects from other incomplete but important maintenance may go unnoticed. Failure maintenance is the glamour job of power generation. PM, however, makes the money. Traditional generation environments promote glamour, not money.
- Light maintenance tasks are worthwhile to do, based on thumb rules, but they don't provide appropriate service intervals. Nominal light maintenance intervals are conservative because little equipment runs continuously—so much is redundant—contrary to what typical vendor-based equipment maintenance intervals presume.

Coal belt replacement

Replacement costs for a 2,600-foot, 60-inch coal belt run around $100,000. Belts moving day tonnage in and out gradually erode, fraying edges from occasional tracking misalignments. Ultimate belt life—like a tire's—depends on fabric substrate or carcass aging. When belt fabric stretches from deterioration, *checking* (a pattern on the rubber belt surface) indicates later belt life stages and signals inevitable impending failure. For standby service, however, an alternative to replacement is to keep the belt as an interim-duty spare, using it during periods when the service belt is damaged, such as during splice tear-out, out-of-service repairs, etc.

The cost calculation involved in swapping a worn-out belt into standby versus replacement depends on continuous-service belt-service life. The typical service life is 6 years with 14 hours of daily service. Total belt aging depends on tonnage moved, flexure, and loading, which causes primary fabric substrate aging (*see* Table 8–2).

Table 8–2 Coal Belt Replacement

Action	Cost ($)	Avoided Cost ($)	Benefit ($)	Cost/Benefit (benefit/cost ratio)
Replace belt	100,000			100,000/200,000
Deferred belt replacement	(operational)	100,000	Time-value of money	
Standby service			100,000 (avoided cost)	
Deferred replacement period (4 years)				
Present value of deferred cost @10%: .6830x100,000	(Operational use of SBY belt for failure)	68,300		
Savings (4 years)		31,700		0.5
	10,000 est			31,700/10,000 = 3.17
Present value Benefit / Cost Discussion			(10,000) Costs associated with using an EOL worn-out spare	Very high at no operational impact

Where installed redundancy is available, a substantial benefit is realized when belt replacement can be deferred. (Of course, the plant eventually has *two* worn out belts, and one must be replaced! This sets up an interim risk—scheduling a short-term belt replacement while the aged back-up belt must carry the load.) Any outcome that risks production loss offers no benefit, but by deferring replacement of a capital-cost item for half the wear-out period, a significant economic benefit results. If this occurs with only moderate plant-risk increase, delaying replacement may be a good bet.

Managing risks represents cost-control's leading edge. Few plants assess this level of equipment strategy complexity.

Condenser condensate alarm checks

Condenser pH monitoring alarms warn when circulating water is out-of-specification pH. Because condenser tube *admiralty* brasses suffer high-temperature pH scaling susceptibility (resulting in high acid/low pH and scaling), condensate and circulating water-monitoring equipment includes instruments and alarms. Instruments, for all practical purposes, follow a classic random-failure pattern; they fail from random events like maintenance or construction (workers disconnecting and failing to reconnect sensors, environmental stresses, and other factors). Assuming random-failure patterns, incorporating redundancy could protect against cost or availability-limiting failures.

When safety isn't an issue, redundancy is not required. However, the consequences of high-acid condition/low pH are grave. Condenser tube leaks require fast unit shutdown to avoid boiler scaling. Boiler and condenser scaling increase tube leaks rates, as well as reduce heat transfer. Severely scaled boilers suffer hydrogen damage, and those operating consequences cause significantly higher unit unreliability due to boiler tube leaks.

Assuring standby alarms are available for pH excursion (making excursions evident to operators) carries the cost of periodic checks to ensure alarm functionality. With various mean time between failure (MTBF) interval assumptions for this random-failure event, simulation yields optimum alarm check intervals. *Interval* is a fraction of the time required for probability of alarm failure, which can be set by specification based upon risk tolerance—perhaps 90% of the interval required for a 30% probability of instrument alarm failures.

Task scheduling for *extremely risky* to *risk averse* failure finding can be selected. Alarm check costs can be reduced even further if they can be built into an operator round, essentially eliminating maintenance support cost/direct entries. Designs strive to incorporate checks into operator rounds and pre-start checklists precisely because they reduce maintenance cost. Operators can perform checks on a tight interval in random-failure cases where periodic testing is required (*see* Fig. 8–3). This is the essence of RCM.

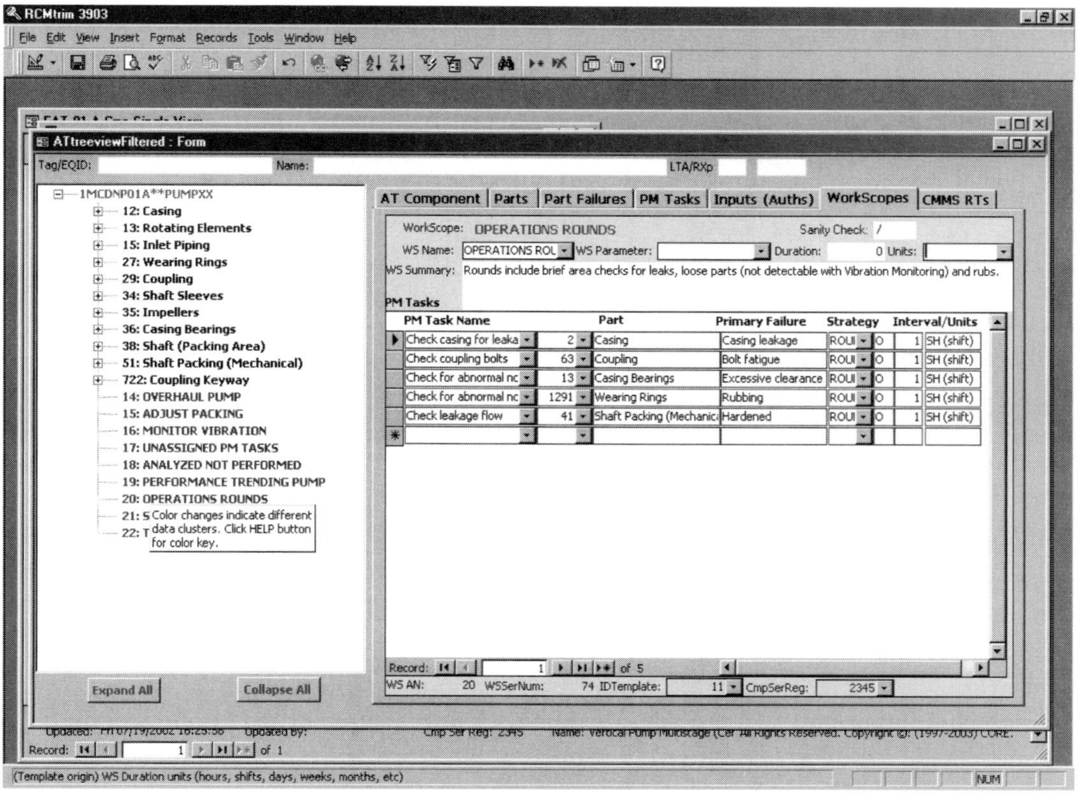

Fig. 8–3 Periodic Operator Checks

Valuable Many

In contrast to a few extremely high-value PM tasks on high-value equipment, many low-to-moderate value tasks on inexpensive equipment can either be performed or the equipment can be treated as run-to-failure. The latter includes area lights and ventilation fans where no immediate consequences occur from failure. The temptation, of course, is to ignore failures that *do* occur until conditions cause secondary failure. (Ventilation fans and lighting are usually highly redundant.)

An example of succumbing to the temptation to ignore failures is a reactor-refueling floor where overhead lighting included 40 high-intensity sodium area lamps, eight of which were suspended below the bridge crane on the bridge. Accessing these lamps for bulb replacement was difficult due to rigging requirements and installation inadequacy to easily access the bridge to replace lamps. Worse, failing lamps could not be replaced on the bridge crane due to safety-related issues. The bridge and all loads were included in the facility FSAR heavy-loads program because the loads were suspended over the reactor. Over a 10-year period, lamps failed until only three bridge lamps remained illuminated. Crews had a practical safety issue moving the bridge, for the area directly below the bridge was dark due to the absence of bridge lighting and shading from the ceiling lamps 80 ft above. Ironically, this is where the predominant amount of over-the-vessel work occurred.

The problem came to a head pragmatically: A survey of plant personnel generated a complaint about working conditions. The responsible parts procurement support group for parts procurement had a substantial backlog of spare part requests to fill. They had declined to take on this work for years! After all, floor work could proceed on even though the area was quite dark. Plant technical specifications did not require any level of illumination, nor should they. Engineering, responsible for parts procurement, had higher priorities than buy sodium lamps (which required a dedication process because of the lack of qualified nuclear vendors). The issue was resolved when the supporting group raised the requested parts procurement priority level to the company's nuclear division vice president. With his interest peaked, new bulbs were ordered promptly.

Reactive environments reward reactive responses. Once that culture is established, expected work performance goes unrecognized. Staff who are recognized and rewarded for performance at minimal competent levels set in place a nasty downward spiral. Culture plays no small role, as reliability influences condition-directed maintenance.

The inability to perform *de facto* on-condition (condition-directed) maintenance constituted *de facto* maintenance program run-to-failure. Engaging RCM will not change this culture and may even exacerbate it. Expectations that run-to-failure somehow allows plant personnel to identify equipment to abandon in place so they won't have to maintain it ever again are mistaken impressions of RCM philosophy. This is the wrong expectation for RCM! Rather, maintenance crews should instead look for better ways to avoid maintenance—like improving equipment-part service lifetimes.

WOs that never expire provide another healthy maintenance program test. In working maintenance environments, aged WO populations show exponential decline over time, and no work order remains very long.

Many small tasks aggregate into substantial work, even though their production impact is nil. Large plants have onsite equipment that includes bulldozers, blades (a bulldozer used to push and pile coal), and backhoes at fossil coal plants, as well as cherry pickers, hoists, and even cranes at larger nuclear facilities. Plant operation vehicles allow site personnel to work around sites that are large enough that some locations are a mile or more from the main building. Most vehicles are redundant and have no operating impact. Routine maintenance on site-vehicle equipment is primarily driven by cost—the cost of avoiding an expensive motor or transmission overhaul on a blade, for example.

Even at this level, PM to CM benefit/cost performance ratio is high to very high and makes a clear case for PM. This validates a general PM selection benefit-to-cost ratio. Another equally important second case is one in which the benefit ratio is low and absolute value large: It's useful to do work where the benefit-to-cost ratio is 1:1, if the savings is $100,000. Some large scope-out work has the potential to fall in this range. The following are general thumb rules:

- Lubrications are generally cost-effective to perform.
- Filter changes likewise should be performed at *appropriate frequencies* (but only the ones that age; many don't).
- Manual valves were installed to aid maintenance, and with no indication of failure, should not require any maintenance other than when something actually fails. (Valves that support online maintenance are the exception to this rule.)
- Manufacturer inspection intervals are over-prescribed.

An exception to this last rule is bearing lubrication on small motors and pumps. These pumps are essentially throwaways. Based on service, value, and the short-use period, there is no value in performing any scheduled work on the equipment. This deviates from the general thumb rule to always perform vendor-recommended lubrication.

Risk

Risk appreciation allows people to make informed operating and maintenance decisions. One limitation to PMO-based scheduled maintenance is that users cannot interpret specific operating or maintenance conditions risk posed to operations, meaning it's not *informed*. Relational RCM, tied to design, is specific and exact; PMO cannot replicate this approach. PMO is one-size-fits-all. Because PMO cannot examine specific equipment in its design context, it is conservative, out of necessity.

Without customization by design and risk, maintenance must be too frequent and contributes to infant mortality, requiring corrective maintenance. Tight overhaul intervals flawed in commercial aerospace overhaul results of JT7 PW turbines in 1958.

They are *still* flawed in industries practicing PMO in 2003! That the military heavily supports RCM analysis and age-exploration in a fundamental way recognizes that with limited maintenance depth and operational expertise (and high turnover), unnecessary exploratory maintenance avoidance is mission-critical to equipment reliability and performance. Industry still struggles with this lesson.

Many heavy industrial processes and hardware never achieve the technological obsolescence of the ubiquitous PC. Paper mills one hundred years old still run; average refinery and power plant age is more than 40 years. Long-term maintenance productivity gains will take a paradigm shift. A maintenance performance leap requires technology to complete the picture.

Maintenance through the middle of the 20th century was crafted, and its delivery of exact solutions to maintenance workers in the field was limited by information technology and expertise that integrated the maintenance and engineering functions. Both limitations are history today! Engineering databases provide software bridges, and RCM provides the fundamental technology.

In the current workforce, aging demographics and accelerated loss of skilled workers—in the absence of substituting less craft-sensitive method—will reintroduce maintenance errors into the workplace. There will simply not be enough of the higher-skilled, experienced workers available to go around. Productivity will have to be sought to offset these losses.

Aging pair strategy

Large equipment has limiting dominant failure modes that drive maintenance performance. Turbine blade fouling that occurs repetitively on some machines represent known failure mechanisms that can be anticipated and planned. Fouling certainty offers hard-time rework interval opportunity.

For large equipment with plant or system support roles, hard time may be disadvantageous. Hard-time overhauls forgo the opportunity to delay a task when the condition isn't present. This inability to schedule "real time" maintenance—maintenance at the time of need—depends on finding suitable condition-monitoring activity. Many intensive, intrusive maintenance tasks offer this opportunity, leading to on-condition maintenance program bases. Achieving on-condition hard-time task substitution develops suitable on-condition predictive tasks that effectively determine equipment condition *and* schedule and perform indicated maintenance at the time of need (P-PF).

Knowing the approximate aging interval for failure and the condition monitoring requirements for the equipment are both necessary and sufficient to pinpoint the appropriate time to perform effective maintenance. Replacing many hard-time tasks is simply a matter of developing the appropriate engineering. Caution is required, however. Until a maintenance program matures so that it can perform on-condition maintenance consistently, owners place their equipment at risk. Management must support the changes, assuring prompt performance of indicated condition-directed maintenance.

9 Data Control

Large Work Group Control

Large groups perform and develop facility work. Work groups need data management controls. From a practical viewpoint, data controls such as spreadsheet documents are difficult to manage in shared-responsibility work groups. Scheduled maintenance plan development involves four primary work groups:

- risk engineers
- maintenance engineers/systems engineers
- craft (electricians, technicians, mechanics)
- scheduler advisor planners

Scheduled maintenance development projects (PM reconstitution, PMO, or RCM) are more difficult to manage as projects than their product's daily use and revision. Large files, with many people reviewing and updating similar records concurrently introduce more chances for data conflict. Daily maintenance use, on the other hand, challenges the commitment to maintain and update records and files, obtain approvals, and follow a prepared process and plan (*see* Fig. 9–1).

Many project processes wither and die after participants move on. Once a project passes into daily PM use, the real test of simplicity, friendliness, and intrinsic value (important in the eyes of the process owners) becomes evident. Processes that need more care and feeding than their output delivers are not worth the effort.

No process is perfect; no software is perfect. Spreadsheets and documents are even less perfect, but open formats mask their imperfection. Large work group control in a document or spreadsheet environment is fraught with consistency conflict and errors.

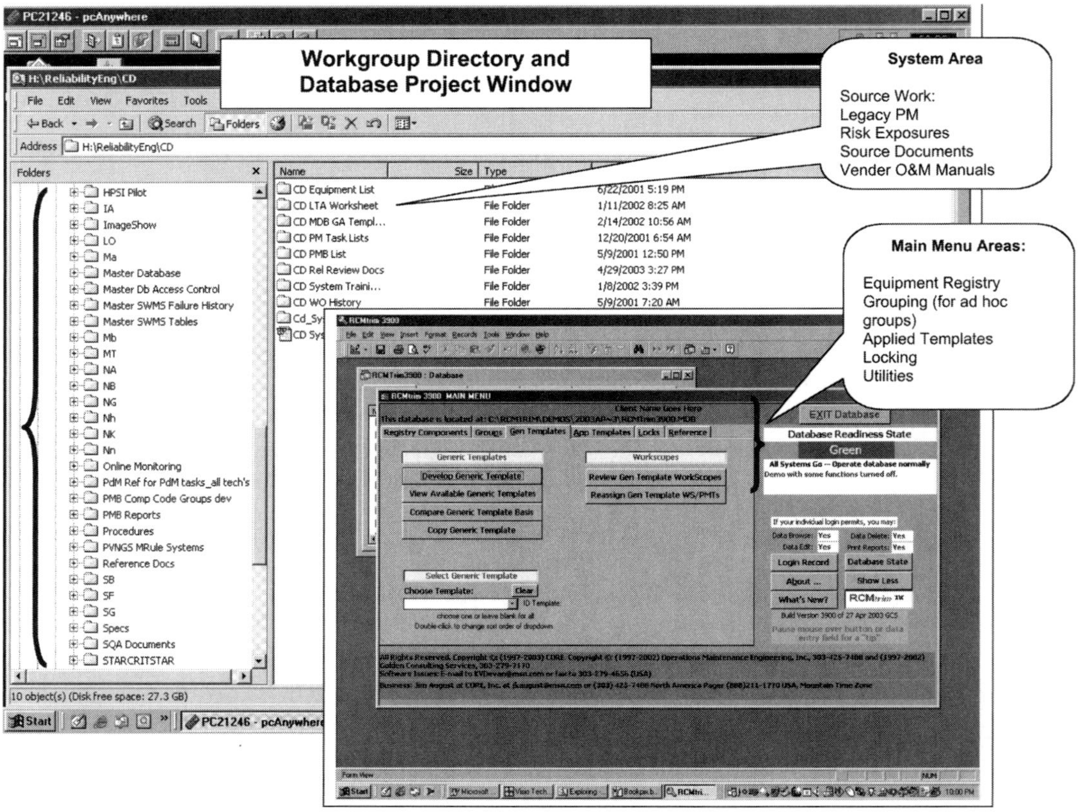

Fig. 9–1 Reliability Disk Workspace

Networked applications offer flexibility, capitalizing on database use of some kind that can realize network potential. That concluded, users and database developmental requirements drive application selection. Document and spreadsheet management improves in-network application, but the intrinsic network group value requires a database to realize.

Work group applications coordinate concurrent users. Automatic data locking prevents data conflicts, a minor inconvenience compared to merging multiple documents or spreadsheets. Large databases—Microsoft Sequel Server and Oracle 9i—can manage hundreds of concurrent users. Large database servers record locking routines, conflict controls, and rollback features in multiple-user environments that make work group data management pleasant. PC applications like Microsoft Access, though not as effective in managing record locks as SQL Server or Oracle 9i, still outperform and out-manage spreadsheets by an order of magnitude. Databases provide system status in larger work group applications. Users with previous experience using desktop MS Office-based Access often prefer it to the server-based database applications (*see* Fig. 9–2).

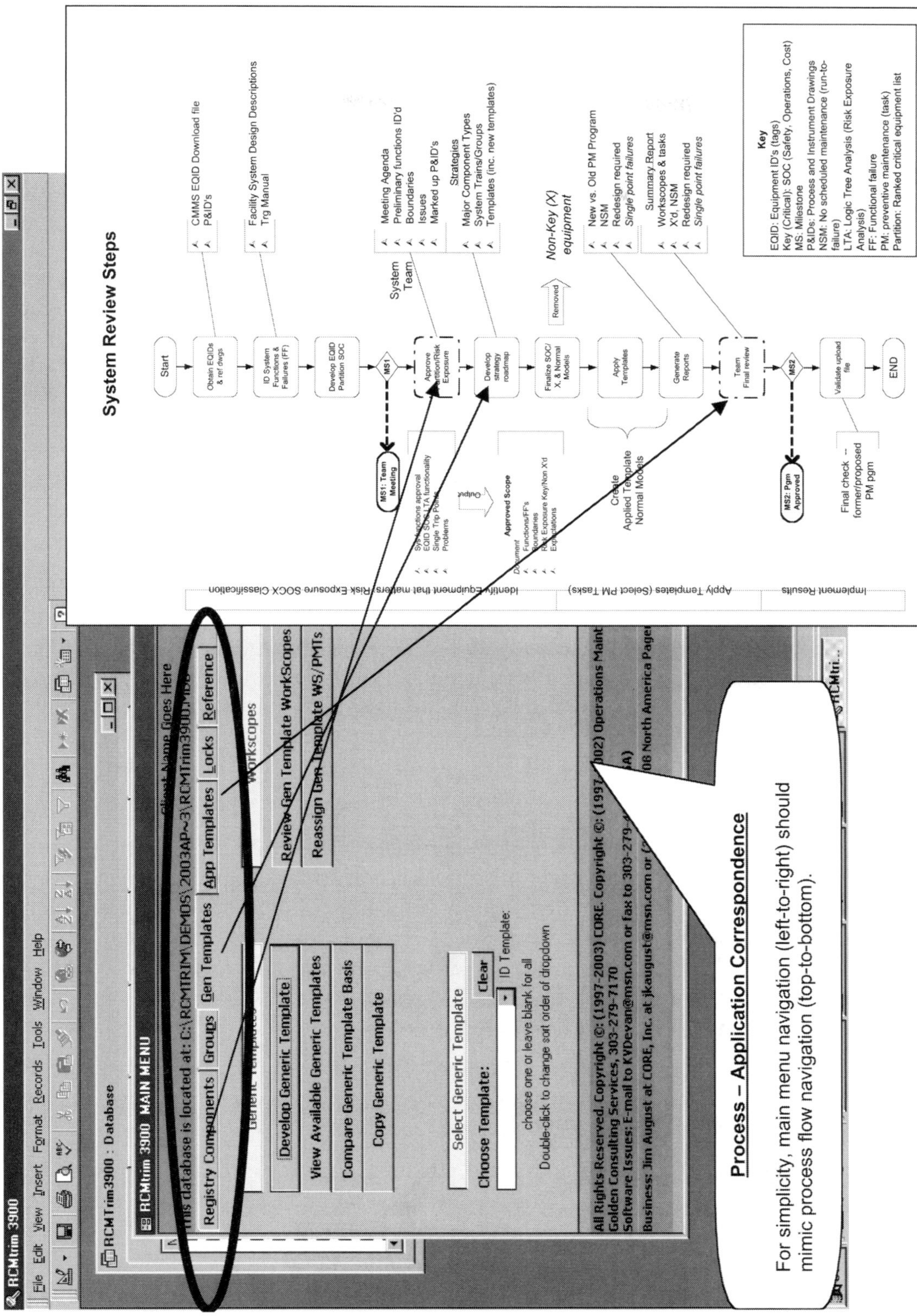

Fig. 9–2 Multiple User Database and Process

Data configuration management

As user numbers increase, controls become more imperative to achieve consistency and minimize data entry locks and frustration.

Database designs can provide some, much or little user entry control, as desired. Databases allow free-spirited expert power users to operate at the periphery of the application's data entry controls. Database boundaries can be much stronger than those provided in either Excel spreadsheet or Word document systems. With a database, users' access may be limited to certain fields for viewing, update, and control. Implementing and maintaining controls increase a database manager's administrative burden.

Multiple users have different areas of interest. Limitations on data changing require customized update authorizations at the user level, restricting the information each can update. Databases track data changes at the record level. This creates a record log—a desirable feature for control. Data changes that reflect final WO workscope tasks and intervals—elements that affect plant equipment PM basis configuration—must be managed.

The value of a database is its ability to manage and track changes all the way to CMMS/EAMS level implementation. Few things are as embarrassing or costly as forced plant shutdowns for regulatory reasons due to scheduled maintenance. These happen where gross lapses on scheduled maintenance oversight occur with consequent equipment failures. The TMI event in 1979 reflected multiple lapses, chained common cause equipment failure, and loss of protective, redundant equipment depth. Although a database would not have solved all TMI problems, better information coordination would have provided a better risk indicator for avoiding problems.

RCM-configured databases will not eliminate chained-failure multiple events entirely, but they provide momentary, historical access to task logic basis. Evidence of work performed is in record history. With a database, risk consequences are evident. Reverse information sorting and diagnostics are easy. (Symptoms can identify likely failures in the field.) Presented with risk information, individuals make responsible, highly informed decisions. Databases track PM task risks and WO resources to manage many potential conflicts and reach optimum solutions.

Change management

The vexing problem in large industrial facilities is tracking complex maintenance requirements for 100 or more systems per unit, several units per plant, and anywhere from 3 to 50 major trains or skids per system with 1000–5000 tags each. That so many analysts do so well is a testament to their skill, fundamental process plant engineering design, and operator and craft knowledge. Yet, it remains problematic that a high percentage of the forced outages that occur have maintenance-preventable causes. The fact that nuclear generation has improved so much, in light of legacy controls and heavy machinery systems in service, has to be that nuclear technologies are mature.

Looking at many of the events that *do* occur, one sees complex processes that fail to convey design requirements clearly to operating and maintenance organizations for implementation. The absence of clear standards (excluding those of the nuclear and aerospace industries) suggests the need for improvement. The relational marriage of design and maintenance with operations in a seamless integrated system is more possible than ever with the new design, logic, controls, and systems available today.

Assuming that the initial plant startup has developed an operating and maintenance plan supporting the design itself, design change control remains. Inexact paper document systems and loose requirement-support ties play a role in many operating events. The seamless integration of these processes is highly desired. Providing methods to integrate design changes into plant operating processes is the first step. All industrial facilities lasting 50 or more years in life experience major changes in equipment and even processes over their operating life. These must be factored in (*see* Fig. 9–3).

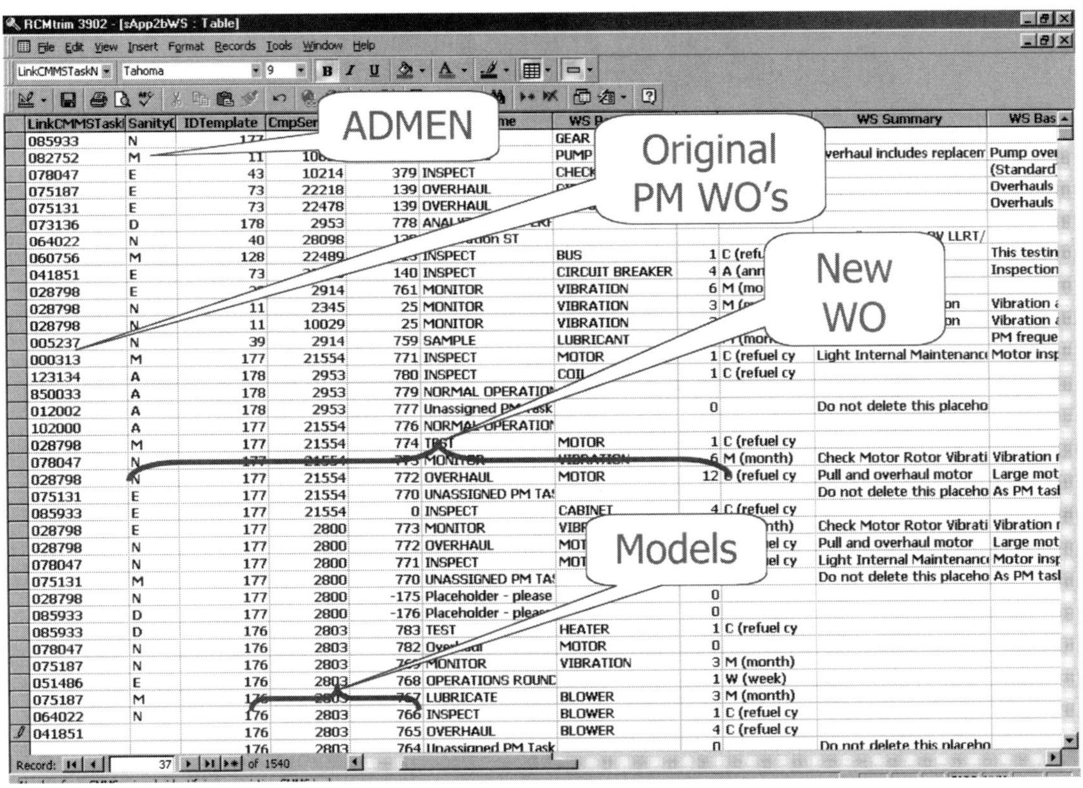

Fig. 9–3 Uploaded PM Changes and Basis Verification and Documentation

10 Standards

Process Standards

Processes center on process standard(s). The first RCM-related process standards address FMEA and root cause analysis and have been available for more than 30 years.

MSG-1 and 2—the first integrated RCM technology guidelines—were developed in the late 1960s. Their current version and successor, MSG-3 version 2 (1993), is still available through the Airline Transport Association. It is mandatory reading for anyone who desires an in-depth mastery of the current RCM practice.

More recently, SAE JA-1011 has emerged to help qualify RCM-based processes. This suggests an ISO 9000 series RCM process certification for companies that provide maintenance services. The existence of a certification program indicates the compelling nature of RCM methods, as well as the inherent difficulty implementing them. Users and developers of RCM-based maintenance programs can evaluate their programs against these standards and reach their own conclusions. For assistance, key ideas in each standard are summarized below.

MSG-3 (2) 1993 Maintenance Program Development Document

The most comprehensive and complete document, the MSG-3(2) Maintenance Program Development Document deserves careful review as the current endorsement of the original works that became RCM. It illustrates the path and complications that arise as considerations for operator monitoring and structural checks get incorporated into the plan. Complexity in the many, highly repetitious flow charts should be studied carefully until the user grasps the concept of hidden failures.

The exact hidden failure terminology—"Does the combination of a hidden failure and one additional failure of a system-related or backup function have an adverse effect on operating safety?"—should be memorized. The user should be able to confidently identify hidden failure from the list, as well as identify failures that probably accompany instrumentation that has hidden failures.

MSG-3 also provides task-selection criteria that select task order. Task selection has been ignored in industrial usage. Its importance is selecting the least expensive tasks to manage failure. The legitimacy of applying multiple tasks on safety-influenced failures should be noted, along with the general caveat to use one task per failure for non-safety ones.

MSG-3 uses language that differs from that used in commercial applications, most significant of which is the reservation of the term *critical* to safety-affecting failure modes. This reflects commercial aerospace use. MSG-3 emphasizes the identification of individual failure modes for task selection—a step commonly lost in other applications.

SAE JA-1011 Evaluation Criteria for Reliability-Centered Maintenance (RCM) Processes

This SAI JA-1011 guideline was developed to address the proliferation of processes that used RCM ideas but deviated from traditional RCM. As such, it was sponsored by interested parties but lacks widespread certification by reliability engineers in forums like the ASME's reliability committee for power groups or the IEEE's equivalent.

Several notable points in this standard are the engineering process focus and its development as a linear-front-to-back model. The process has an excellent set of definitions and provides useful materials to assess RCM methods in other processes.

INPO AP-913 Equipment Reliability Process Description

The INPO has developed a high-level standard that integrates all aspects of industrial plant applications. This standard addresses many implementation issues that are discussed at length in the original Nowlan and Heap RCM analysis washed over in later documents that focus on RCM engineering development.

INPO AP-913 makes points that conflict with traditional RCM development. Most notably, it includes feedback correction loops for the maintenance plan and provides for the development of standard PM plans using templates. Templates are a significant strategy that should be carefully evaluated by large maintenance program users. Their pros and cons are noteworthy and provoke lively discussion whenever engineers meet. Template utility in industry should be obvious.

INPO AP-913 implicitly endorses the NRC's Maintenance Rule 10CFR50.65, which calls for performance standards and corrective action based on two broad performance measures—system availability and maintenance-preventable function failures for covered equipment. NRC and INPO rule performance measurement aspects are important because they are unique and are not addressed in other rules or standards at the depth covered here.

Users may also consider referring to NERC guidelines for performance measurement criteria.

The single greatest difference between RCM and AP-913 is the absence of direct failure criteria in evaluation of critical failures. AP-913 allows fully redundant systems to be elevated to *direct safety* failure rank, in contradiction to MSG-3. By redefining functional requirements to include two safety redundancy layers, the two become compatible.

MIL STD 2173 Reliability-Centered Maintenance Requirements for Naval Aircraft

This standard provides the Navy adaptation of MSG-3 and RCM for Naval Air applications. Its main value is to show service endorsement of RCM, and how that is implemented. This is a comprehensive, yet directive approach towards implementation.

Reliability Centered Maintenance by S. Nowlan & H. Heap

Any discussion of standards must include Nowlan and Heap's monumental work, if for no other reason than it established the RCM genre. Where questions of legitimacy or validity come up, it's appropriate to review the originator's work to see how they addressed the same concerns. There are few omissions or oversights in this monumental work.

Unfortunately, this document has been out of print for at least a decade, even though it remains the definitive work and an outstanding reference when questions of interpretation arise. In the past few years, several secondary reprints have been available. Anyone seriously pursuing RCM will want to obtain a copy of this work.

11 Software Applications

Software

RCM has been performed in software text and spreadsheet forms for more than 20 years. Software still offers an opportunity to streamline RCM analysis.

Early RCM editions provided excellent work for their day and remain excellent methods for focused, detailed component analysis in rote detail. However, larger projects, better software platforms—especially databases—and more complex industrial users make new demands requiring new perspectives.

New databases speed analysis when compared to their mainframe antecedents. PC databases are over a decade old; some are obsolete! New network and web-friendly databases are available with advances in server technology for Internet use. These provide exciting new platforms for RCM technology development.

There are at least 20 commercially available RCM database software products today. (The author is associated with one.) These offer services that range from teaching RCM to performing detailed classical RCM required by MSG-3 to parts certification. Some of the companies also offer middleware products that interface with CMMS/EAMS. Other companies document technical bases that augment RCM analysis. Companies considering RCM implementation should seek and examine alternative software vendors.

Occasionally, operating companies develop their own commercial products. Large companies have considerable resources. Startups in contrast offer innovative approaches. Some companies' strategic business plans can afford financially to support new software development. Most cannot. RCM software presents classic make/buy decisions. With the available products, companies should ask whether they could improve upon commercially available products that enjoy wider user bases and acceptance.

Productivity and speed

Improved productivity results from RCM software. Among the early power generation CMMS/EAMS software applications developed in the 1980s, few remain. Software evolves fast. In contrast, many similar or identical equipment applications are still in service. Many systems have multiple identical trains that lend themselves to exact analysis replication. Many common production components (such as large high-pressure steam service safety valves) are extremely similar, even when multiple supplier products are considered. These installation constants make RCM application software even more beneficial.

Users applied design symmetry and similarity in early CMMS (known then as maintenance information systems or MIS) to develop PM models. Early software would not support *siege analysis* techniques so large projects with hundreds of thousands of components—whole power plants!—bogged down due to software problems and scrambled spreadsheets and generally could not be efficiently managed. Analytical tools spawned text document formats authored in early applications like ATMS. From these early applications, Lotus 123 spreadsheets and other PC-based products evolved.

System users experienced extreme frustration with software limitations. Alternative products offered few efficient ways to use design similarity, symmetry, and replication to simplify work. Ultimately, finished PM model work orders in intermediate form, provided as templates for repeated PM work-order application as PMs came due. Working directly in CMMS/EAMS products like Power Plant Maintenance Information System (PPMIS), an early non-relational plant maintenance software product) was tedious. The absence of text editors made even the simplest task like text wrapping a tedious demanding chore. No modern software user would tolerate that today.

The latest CMMS/EAMS applications provide much better work-order development and maintenance processes, but none offer elegant RCM-based reliability solutions with extensive template-based, basis-maintaining PM development capability.

Objectives

For commercial success, RCM applications must offer productivity and make users happy. An accepted IT axiom is that software users show no mercy. They fault software for performance, regardless of cause. Hardware, systems, resources, or other issues do not matter to users. They view problems simply as "The workstation doesn't perform!" and software is the cause.

Whether the network has adequate server support, sufficient processors speed, large enough hard drives, adequate cable capacity, or adequate RAID levels are immaterial—performance gets attributed to the application.

Software owners (IT and engineering service groups) that make software packages available for use must understand and anticipate their organization's users and demands. Browsers, data entry users, and batch processors all must be reasonably accommodated. Software that only works well for a small engineering PM development group won't work well when operators browse to diagnose equipment. While some obvious reasons to automate RCM processes are development productivity and speed, other uses and users are important to recognize. Maintenance browsers need report capability; management wants statistics and traceability of bases. When a system designed for five concurrent users suddenly supports 25 or more, performance issues abound. Under such circumstances, RCM software objectives could include

- develop PM tasks and intervals
- develop PM tasks and intervals, with a basis
- develop PM tasks and intervals, and facilitate their planning into workscopes
- develop PM tasks, intervals, and basis and allow batch upload to the CMMS/EAMS
- develop PM tasks, intervals, and basis and allow batch upload to the CMMS/EAMS with justification auditing of all changes
- provide a way to maintain a regulatory PM basis[1]
- provide a way to maintain the engineering basis[2]
- provide operators with real-time diagnostic tools
- provide work order performers risk information about the equipment in question
- provide scheduled maintenance program performance monitoring information
- identify maintenance resource allocation by equipment or risk classes
- identify groups of equipment with similar risk, regulatory, or other attribute tied to scheduled maintenance
- demonstrate compliance with codes and laws
- maintain a living maintenance program
- document all known dominant failure modes in a facility
- relate dominant failure modes to hardware over a facility for risk management
- develop statistical reports for scheduled maintenance strategy distribution
- document the available site skills repertoire for performing PM

What started out as modest engineering software suddenly blossomed into a full-fledged database with many users and major interfaces with other software like the CMMS/EAMS, with different users finding different needs and interests.

Although primarily browsers, operators print reports for rounds. Work planners re-plan work orders to develop workscopes—which they historically planned into WOs work package. Reliability engineers ensure that critical-equipment PM is approved for work that is specified, planned, and performed as scheduled. Managers seek top reliability risks and cost reports. Work control seeks uploaded batch files that provide outage tasks and extensions 12 weeks ahead of outages so that scheduling can be regrouped.

Aside from practical objectives, broader, philosophical questions influence any software design.

For example: A trade show attendee asked a vendor whether their product performed traditional RCM. The response was, "It can," to which the questioner's retort was, "How can you call a software RCM-based if it allows users to perform anything but faithful, traditional RCM?"

Imagine asking Microsoft the same question of Word and the corresponding response, "How can you call MS Word *word processor software* if the user can crank out poorly formatted documents or embed spreadsheets or add pictures?"

Open software architectures enable, but without controlling the customer's end use of the product. Experienced engineers as users, agents, and developers acknowledge that there are two broad software design philosophies. One provides tools with few restrictions that enable experts as well as neophytes. Software provides limited controls, but doesn't restrict expert use. The second design philosophy type provides complete control; it elicits very specific responses from users based on a series of questions, field restrictions, and interactions. In responding to certain questions, certain pathways open up one set of options; another response opens a second set.

All software contains elements of *both* design philosophies. Practically, *open* architecture designs appeal to expert users, as they are more flexible and allow more creative freedom, faster work, and fewer user restrictions. System constraints insult expert users.

Application software forces the user into the application environment. Opening Word, the user works in a Word environment with Word terminology, formatting, features, and controls. Clicking Save generates a Word document (*.doc) that is of little use in another product like DB4! This specificity is the attraction and damnation of any software.

Working in an obscure product results in the limited use of the work by others. To avoid this, engineers often work in very-common-and-accepted MS Excel spreadsheets.

Spreadsheets are attractive when the user isn't exactly sure how to proceed. For software design development, Excel works well to draft rough relationships and build sample data. Using another application or saving Excel data to another application's format restricts the use of that data to the application. For those used to spreadsheet flexibility, controls imposed by a database application make life difficult. They can't immediately create a new field, enter data of their choosing, or perform drag and drop fills (as in Excel), copy/paste sheets of data, etc.

Some people struggle to make application transition, resisting the uniformity imposed by the application. Some engineers are independent *prima donnas*—plant engineers are worst of all—but uniformity is why an application is necessary for a workgroup. Try doing a project with 10 people in a spreadsheet.

Customization vs. control requires balance. An outstanding product walks this line carefully, keeping users happy. User experience ultimately determines market acceptance. Companies that anticipate software purchases are well advised to survey proposed product users to gauge their acceptance.

Customization

Large application software requires user acceptance if it is to provide maximum benefits. For large software applications, the greatest efficiency occurs when RCM data and results can be used by other plant software systems. The CMMS/EAMS system and material control applications are the closest potential interfaces. Meeting this interfacing capability requirement necessitates *middleware*—a software interface between the RCM application and other systems, which requires customization.

Customization can be beneficial or deadweight. It's expensive, lucrative for developers, dear to clients. It has been called, "the gift that keeps on giving." Development fees are substantial, but purchasers are obliged to fund some customization for their organizations if only to gain installation acceptance.

Customization can easily grow out of control. Given an unlimited installation budget, software users seek many software changes, making many demands and requirements. These demands can cross the boundary from value-added to specious. In extreme cases, change requests become circular between the needs of different work groups (or even the same work group over time).

Organizations procuring commercial software seek maximum customization with the developer absorbing the cost. After all, why should they pay customization expenses for features that will add value to the base product for the software company's next sale? Unfortunately, user-demanded customizations are rarely marketable. Users demand features that they believe others need, but others have their *own* sets of needs—and will only pay for them.

Customization can be a ploy used to seek discounts from developers. Paying real costs for customization quenches user demands quickly.

Process connectivity

Integrated software drives organizational processes. Software processes can be accepted or contended. Engineering firms in which members historically have had many degrees of freedom react more coolly to software constraints. Controls that are inconsequential for clerks threaten engineers. For organizations, software processes must truly reflect organization needs and processes.

Compared to electric utilities, other industries have less structure. Their processes and core strategies haven't been as fully defined so users are more flexible. The nuclear electric generation industry carries structure to its limits. Vendors try to understand their clients' business needs and define goals around them. Identifying and implementing process software requires interpreting those needs. Interactions to develop business processes and corresponding software generate stress, but software that enhances productivity will reduce stress over the long haul. Software that is burdensome, troublesome, or difficult to maintain will not be acceptable in the long term. Achieving software development and implementation teamwork is difficult, but when achieved, win-win outcome potential is high

Some software is just not useful. It can constrain, demand, be inadequate, buggy, or just flat out fail. (One engineer coined the term *feeding the software* to define his perception of how software becomes an end in itself.) It is an especially unfortunate situation for the engineer who inherits products like these. Sometimes upper management makes purchases based upon previous work experience with a particular vendor without the involvement of the user community. Software *caveat emptor* applies. Users should fully test new software products and test features *before* they buy. All subroutines should be examined, performance times tested, and prospective user acceptance sought. Also organization learning curves should be considered.

Given these precautionary measures, there's still no guarantee that new software will make an older version or product obsolete. One need only look at the history of software evolution over the past 20 years to see this. Users should strive to understand basic software models, paradigms, and processes to ensure that the products they procure meet their needs.

Completeness

Software-based RCM stems from the need for process completeness and integration. RCM is complex; software-based RCM ensures that easily overlooked points are addressed. Spreadsheet analysis cannot provide these assurances. For analytically complete answers, there is no substitute for software that makes certain that a process is followed and brings omissions front and center and delivers consistent products.

Technically comprehensive, complete answers makes a strong case for software that leaves no stone unturned.

---

[1] The regulatory basis has compliance demonstration as its primary objective.

[2] The engineering basis focuses on providing the reliability engineering rationale behind each PM task.

12 Conclusions

Performing RCM in industrial applications can be reduced to doing three basic activities very well:

- developing an equipment risk partition
- applying templates to the critical components in the partition
- uploading the results

This process is simple enough to allow knowledgeable plant personnel to perform RCM analysis well with proper administrative support and control. To rise above historically inexact, *ad hoc*-decision, craft-based maintenance processes must focus on process development and control. This methodology challenges traditional separation of maintenance, engineering, and operations. Developing a successful RCM process promotes teamwork that reduces that separation.

A successful process will very likely require RCM databases, interface middleware, and implementation process development into the CMMS/EAMS.

Historically, RCM could not easily span many required technological elements to be cost effective and successful. RCM technology wasn't clearly and sufficiently understood; software tools like local area networks and databases that facilitate information movement weren't advanced.

Those barriers are gone today. World class organizations can't afford to not consider RCM seriously in their maintenance programs today.

Glossary

Advanced Text Modification System (ATMS). A 1980s mainframe computer text editing software in widespread use at that time.

Age exploration. Systematic, planned examination of the most aged members of an installed in service population to validate expected aging performance.

AP-913. Equipment Reliability Process Description (INPO).

Applicable and effective. Technically correct and effective (e.g., works in a real production environment) and cost effective to perform compared to acceptable alternatives. Acceptability weighs social and cultural values including value of life, and the environment.

Applications. A component applied template or software.

Applied template. The applied counterpart to a generic template, where only expressed failure modes are used (applied) from a smorgasbord of alternative selections. The generic template is like a crime composite; it reflects all possible failures averaged into a common melting pot. The applied template seeks to extract and refine details again.

Applied templates. The various features of a component generic template model that have been selected to apply to a component tag number in a real plant creating a real equipment context.

Appropriate frequencies. Frequencies that reflect service and failure mechanism. These differ from manufacturer recommended frequencies primarily in that they operate only part time, on average, while manufacturers assume they run full time.

As builts (as built drawings). The final plant construction drawings that summarize how the plant was actually physically completed.

Associated equipment. An equipment tag in a primary association equipment group. For example, in an instrument loop, a transmitter may serve as the tag. All other components in the loop would be under the transmitter tag.

ATMS. See Advanced Text Modification System.

At-risk (failure). Place where a failure has a high probability of occurrence based on the benchmark assessment of conditions known to contribute to failure events.

Blocking (task blocking). Organizing tasks into logical WO blocks of activity for efficient work performance based upon skill required, engineering interval applied, tagout boundary, plant operating mode to perform the tasks, and other more subtle factors. Task blocking usually requires shop, operations, and engineering input to optimize around many constraints.

Boiler and Pressures Vessel Code ("The Code"), Section III or X. The ASME's certified code for managing the design and maintenance of pressure vessels and steam boilers for power plant use. Initiated about 100 years ago to control design and operation of boilers to avoid explosions, the code has gradually evolved to include pressure piping and large pressure vessels like nuclear reactors. The code is usually cited by Section, which applies by class to various industry segments. Section III applies to nuclear pressure vessels, for example; Section X applies to unfired pressure vessels.

Bridge crane. An overhead crane configured like a bridge spanning the walls of a building.

Bridge catabase. Middleware software that bridges from the RCM application or output tables into the CMMS/EAMS application tables.

Buna-N. Buna nitrile. An oily synthetic rubber still in common use for o-rings and other common plant elastomers. Black rubber.

Buyer (purchaser). A person who buys services and materials for an industrial facility.

Carcass. The fabric backbone of the belt or tire impregnated with rubber or other elastomer. The fabric substrate or web that provides structural support.

Check valve. A valve that checks flow in one direction preventing reverse flow.

Code case. A proceeding brought before a code-certifying committee to change the language or interpretation of a code—like the ASME Boiler and Pressure Vessel Code.

Codes. Standards that are endorsed legally to carry the force of law. The ASME's Boiler and Pressure Vessel Code, ("The Code") is the prime example. Many laws, such as the U.S. NRC's 10CFR50 simply refer to the code for technical compliance requirements.

Cohort. The users of a common applied template normal model.

Complex equipment. Equipment that never exhibits dominant failure modes other than random failures. Equipment that empirically lacks any predominant age failure characteristic.

Component identification code. Component tag number.

Condition-directed maintenance. Maintenance directed (e.g., scheduled) to be performed based upon a condition identified. The performance of a scheduled maintenance task conditionally initiates the actual hardware maintenance. Although superficially this seems to be the same as corrective maintenance, the subtle distinction is that the maintenance restoration work is triggered by a scheduled activity—a check inspection, or test, that identifies (potentially) the condition that requires the maintenance. (Example. An annual turbine test may detect performance off the thermal efficiency curve; this condition would initiate a condition-directed rework of the turbine blading to restore efficiency once the performance limit had been exceeded.)

Core damage (reactor core damage). Damage from inadequate cooling and excessive temperatures from high heating creating local cladding weaknesses, releasing radioactive fission products into the cooling water. This reflects a design barrier lost.

Data (real data). Actual collected parameter values from data logging systems and field measurements that provides source material for condition assessment and analysis.

DB4. An early PC text editor no longer in widespread use.

Dedication. A process whereby commercial grade equipment can be specified and used for nuclear safety applications by ensuring the materials will meet performance expectation and providing a lineage.

Deterministic. Certain, derivable; the opposite of probabilistic.

DFM. Dominant failure mode.

Dispersion. Spread about a mean. Variation in characteristics.

Drift. Gradual deviation from a set adjustment.

Drylab. Performed on paper only. Other synonyms included *gundecked* (Navy), *radioed, faked,* and *cooked* (cooked data or books).

Effects (failure effects). Local and chained effects of failure. Local effects are the immediate or proximate effects of failure. Failure chaining requires an understanding of the logical connections of the equipment in the systems of interest. Local effects may be incorporated into templates; chained derivative effects require system analysis (fault trees).

Elastomer. Rubber-like. Made with organic materials and cross-linked compounds.

Engineering failures. Failures defined in fundamental physical process terms such as abrasion, stress corrosion cracking, or Arhenius temperature aging.

EQID. Equipment identifier.

Equipment identifier. Equipment tag number.

Erosion/corrosion (in high energy piping). Accelerated loss of metal from a loss of protective hard oxide layer deposited as a result of the high-temperature corrosion process in new plants. Loss of the hard layer accelerates wall thinning, which eventually leads to line rupture.

Errors of commission. Errors that include tasks erroneously or that add work that can't be firmly justified upon closer examination.

Errors of omission. Oversights, missed calls.

Evident (failure). A failure normally evident to the operating crew.

Exposure. Risk exposure—the specific failure risk quantified in safety, operational (production), and cost (SOC) terms by using a task selection logic like that found in MSG-3, SAE JA-1011, or another logic risk tree assessment process. Usually codified mnemonically into *S*, *O*, and *C* classes.

Fails to open. Sudden failure on demand due to an open loop condition caused by a failure in a loop component.

Failure effects. *See* **Effects.**

Failure modes and effects analysis (FMEA). The qualitative analysis of likely equipment failure modes, and their effects—local and otherwise. Likely failure modes restrict focus to events that credibly can happen. Statistically, the determination that 93% of all known equipment failure modes are random or occur outside the economic lifetime of the equipment in question must be factored into the equation. This ensures that resulting analysis is relevant to the equipment in question rather than being an academic exercise in enumeration.

Failure modes and effects criticality analysis. Similar to FMEA but including criticality factor calculation for each failure mode identified.

Federal Energy Regulatory Commission (FERC). The U.S. federal commission responsible for the regulation of interstate electricity sales. Because electricity sales on the grid are inseparable intrastate to interstate, FERC requirements practically extend to all companies selling wholesale electricity into the grid. FERC requires cost reporting for power generation facility costs for regulatory compliance purposes. Although full of reporting errors, FERC cost databases provide a means to independently validate failure data based upon cost against NERC data.

FERC. *See* **Federal Energy Regulatory Commission.**

Five causes. *Total Quality Assurance, the Japanese Way* cites the five causes for any problem. The goal is to simply ask, at least five times, "Why." In doing so we should reach the root cause for a problem.

For lack of a nail. (traditional, cited by Lynch and Kordis, 1992)

For lack of a nail, the horseshoe was lost,
For lack of a horseshoe, the horse was lost,
For lack of a horse, the rider was lost,
For lack of a rider, the battle was lost,
For lack of a battle, the kingdom was lost.

Five causes of failure. For any failure, to ensure tracing the root cause back to the source, the failure investigator should ask "why?" a failure occurs at least five times to assure themselves satisfactory root cause analysis (from Qaizon. *Japanese Total Quality Control*).

Fleet leader (fleet aging leader). The equipment in a fleet with the most service and aging cycles based upon use.

FMEA. *See* **Failure modes and effects analysis.**

Force measure. Something with amplification effects; more than on the face of it.

Final safety analysis report (FSAR). An assessment of a nuclear facility design that provides a key milestone to go ahead with construction.

FSAR. *See* **Final safety analysis report.**

Fubar (foobar). Fouled up beyond all recognition; hosed.

***Gatronix* public address systems.** The primary supplier of plant PA phones and systems. The trade name *Gatronix* has become synonymous with the product public address phones like Kleenex and facial tissue.

Generic letter (GL). An NRC letter to the industry documenting a problem and recommended actions.

Generic template. The source model for developing equipment.

GL. *See* **Generic letter.**

GL 89-10. Motor Operated Valve Problems.

Going down. Tripping offline from protective devices or operator action; an unplanned forced outage.

HE. *See* **High energy.**

Heavy loads program. Nuclear plant special programs to lift and move loads over safety-related nuclear equipment and the reactor itself.

Hidden (failure). A failure normally hidden from the operating crew.

Hidden. Not apparent to an average operator performing normal duties.

High energy (HE). Locations where high temperature and pressure steam is contained.

High energy steam piping. *See* **High energy.**

Hope-for-the-best. An inexact, wishful way to seek improvements. Lick and a promise, etc. Quickly done, with more wishing than analysis.

Implement. *See* **Operationalize.**

I&C. *See* **Instrumentation & control.**

In context. Considering the risk, service and environmental factors that influence the dominant failure mechanisms expressed and the risk posed, which combined determine the best PM task strategy for controlling risk.

In the equipment. People who work directly on the equipment and hardware, whose hands are "in the equipment."

Infant mortality. The observation that in an aging population following a general Weibull distribution, upon initial use a high failure rate gradually stabilizes to yield a hardened population of viable equipment that exhibits low random failure over its lifetime. These results were first discovered studying human population mortality. The phenomenon was first described as "infant mortality."

INPO. *See* **Institute of Nuclear Plant Operations.**

Institute of Nuclear Plant Operations (INPO). A nuclear trade group with mandatory participation for companies operating power reactors.

Instrumentation & control (I&C). A craft work specialty that maintains electronic instruments and controls.

ISO 9000. A European Common Market standard (series of standards) that ensure companies have developed processes for quality in the manufacture and delivery of products or services. These standards ensure processes are mapped so that the control of variations and defects in goods and products should be controlled. Buyers of services from ISO 9000+ certified companies have assurance that these companies control their work processes for quality products and services delivered to customers. ISO 9000 is transparent to the company's product or processes; it merely ensures they are controlled. As a process standard, it's like INPO AP-913, only more general.

Iteration. One stage of iterative changes; one rework step.

Iterative. Repetitive incremental adjustments reflecting rework.

Law of unintended consequences. Doing complex things produces the opposite outcomes form those intended. Outcomes cause consequences opposite to those that are intended or action taken could yield unexpected results.

Leading age. First in aging experience; accelerated aging.

Leading age group. A group of equipment in service with more run time and aging accumulation than the rest of the fleet.

Lift-off tests. Tests that raise the valve off the seat to demonstrate freedom of travel, flow pathway obstruction, and freedom of lift device for critical safety relief valves.

Limiting DFM. The most restrictive dominant failure mode of many that might drive an overhaul. The one that forces the maintenance work order to be scheduled on its failure interval.

Living maintenance. An ongoing actively evolving maintenance program, changing with new methods, technology, and equipment performance findings. A dynamic engaged program.

Local effects. Effects in the immediate proximity of a hardware failure, or affecting immediate functional requirements for the equipment that suffers failure; not remote or secondary. (Secondary effects may be distant, indirect, and reflect integration.)

Logic tree analysis. *See* **Task selection logic.**

LTA. Logic tree analysis.

Master equipment list (MEL). The design equipment list or the registry.

Mean life. Average life or age in a statistical population.

MEL. *See* **Master equipment list.**

Middleware. Software between other software systems, often to complete business loops that one or another software doesn't complete.

Missiles. Projectiles from rotating equipment failure with high rotational inertia parts. As parts fail due to inertial forces, they become ejected missiles.

N-dimensional. Alluding to the multi-dimensional characteristics of advanced mathematics, which also has the unintended side effect of confusing the user.

NERC. *See* **North American Electric Reliability Council.**

No-load. Unloaded, without a coupled load.

Normal model. An equipment tag or number that provides a modeled component template used on other similar equipment with identical context. The equipment identified from the plant master equipment list (registry) that uniquely identifies the applied template for a class of essentially identical equipment in identical service. Also called master model

North American Electric Reliability Council (NERC). A voluntary industry group formed to ensure electric transmission system reliability. Members must report failure data to FERC for reliability analysis on a real-time basis; many non-members also report as a matter of practice. NERC reliability statistics may be obtained anonymously at cost as reported for member plants.

NSM. No scheduled maintenance.

O&M. Operations and Maintenance.

Object database compliant (ODBC). Databases that meet a standard so that their tables can be manipulated by any other compliant database. A database connectivity standard that ensures the ability to access and manipulate data no matter what the users' interface software is.

ODBC. *See* **Object database compliant.**

Operating year. One year in operation, which may be many calendar years for partially-run equipment.

Operationalize (implement). Complex laws and rules like the 1992 American Disabilities Act are not straightforward to implement. Operationalize means figuring out how to implement something such as this.

Over-select. Include more failure modes than conditions or experience suggest are dominant for a piece of equipment. Inclusion of rare or unexpressed modes kluges up a program with unnecessary inapplicable tasks.

P&ID. *See* **Process and instrumentation drawing.**

Pad weld. A pad of weld material laid over a weak area such as a boiler tube leak. An inexpensive but impermanent weld repair.

PID. *See* **Process and instrumentation drawing.**

PF-F. *See* **Potential failure to failure.**

P-PF-PMT. Part-Part Failure-PM Task.

PM models. PM WO work plans ready to be attached to routine PM work orders. Essentially, before electronic work control systems, these were copies of standard plans to attach to control document for issue to work. Today the copying and pasting into control documents is electronic although the principles are the same.

PM optimization (PMO). A simple PM Optimization approach that uses templates by equipment type to develop customized maintenance plans. PMO's most significant difference from RCM is the absence of risk consideration in the task selection process. PMO has been performed in power generation since the early 1980s and represented state of the art PM program developments when first introduced in text (DB4, WordPerfect and Word) and spreadsheet (Lotus 123 and Excel) formats.

PMO. *See* **PM optimization.**

Potential failure to failure (PF-F). The time from failure detection to full expression. Incipient to mature failure development period.

Powder River Basin (PRB) coal. Coal from the Powder River Basin of Wyoming with common combustion, impurity, and firing characteristics. A very low heat content, volatile sub-bituminous coal popular for low sulfur and cost.

PRB. *See* **Powder River Basin coal.**

Primary key. A unique identifier like a social security number used in databases to track unique equipment and records.

Primary tag. A component tag that uniquely identifies a primary secondary function group. The primary tag can trace performance.

Process and instrumentation drawing (PID and P&ID). The fundamental design drawings for process facilities.

Protected failure. A failure precluded by an interlock or made evident to an operator for action by warning alarms, trips, or enunciator. A failure protected by an alert, trip, or other alert control function that identifies and precludes the otherwise hidden failure.

Quality assurance (QA). A formal department charged with maintaining quality standards and services. A quality oversight group at nuclear power plants.

Random. Occurring unpredictably, except for frequency of occurrence. Opposite of certain or known.

RCM. Reliability-centered maintenance.

Reactor core damage. *See* **Core damage.**

Real Data. *See* **Data.**

Reconstituted. Redeveloped, particularly supporting analysis.

Record strings. The selections of Part-Part Failure-PM Task (P-PF-PMT) that uniquely define one applied template PM task activity by part and failure.

Redundancy. Deliberate duplication or partial duplication of system components to decrease the probability of a system failure.

Redundant. Equipment provided for redundancy purposes.

Refueling floor. Plant level or floor in a nuclear facility from where nuclear fuel is loaded removed or otherwise manipulated. Area above the reactor top head and spent fuel pool.

Retrievability. Ability to recover and examine, usually for source engineering, design, construction, or manufacturer documents.

Rework/replace. Hard time time-based maintenance.

Riders (travelers). Documents that accompany a work order like schematics, drawings, procedures, or vendor O&M excerpts.

Root cause. A cause that, if eliminated, prevents recurrence of an event.

Root cause failure analysis (RCFA). A methodology for identification and elimination of root causes.

Run-to-failure (RTF). No scheduled maintenance (NSM.)

Safe life limit. A conservative part failure lifetime which ensures 100% of aging parts are replaced prior to failure based upon the safety consequences of the failure mode. A hard time age limit for the conservative replacement of a known aging part before *any* failures can result. (For example, if we considered an airplane's landing gear tires safety-based—which was the case for the Concorde—a safe life limit might replace those at 100 takeoff landing cycles, even though the mean life was 500. We want 100% assurance no tire fails in service due to age!)

Sanity check. A final check in the CMMS/EAMS PM schedule subroutine table inputs to confirm that changes to the PM WO workscope tasks, descriptions, and intervals are exactly correct and ready to be loaded into the production database.

Showstopper. Barrier to production design licensing. Something that must be addressed favorably to move forward.

Siege session. A closed-door, work to completion session to complete a task quickly. A focused intense session intended to draw closure prior to dismissing participants.

Silver bullet. A quick, simple fix. The ideal solution to any problem—evident, simple, and cheap—and for this reason, oftentimes not available in the real world.

SOC. Safety, operations, and cost.

Sootblowers. Blowers (long retractable air jets that blow a variety of gases [air, steam]) to remove soot from boiler tubes. Soot removal maintains boiler efficiency and temperature pressure relationships, keeping the fire and steam phases correct in the various sections (waterwall, superheater, reheater, economizer, air heater, etc.) of the boiler.

SRCM. *See* **Streamlined RCM.**

Streamlined RCM (SRCM). Abbreviated forms of performing RCM. The reader is cautioned to consult standards like SAE JA 1011 to assure them that SRCM meets their needs.

Status only. Instrumentation that has no specific failure alert, trip, or control function or general function within an active control loop on critical plant equipment. The role of the equipment is to provide status, and it typically has multiple other redundant means to obtain status from other available instrumentation—or the status information provided is for non-operational purposes (like startup).

Strategy (in a PM Sense). One of four basic scheduled maintenance options: time-based maintenance (rework/replace), condition-monitoring (predictive), failure finding (discrete), and none (no scheduled maintenance). One can iteratively add to this list, with redesign or trending. There aren't any other basic options to performing scheduled maintenance.

Tag number. Also, tag no. Equipment tag number.

Takeoff or takeoff list. A list of work taken off the P&ID or other approved summary of the equipment required in a system and its process and physical relationship to the plant.

Task selection logic (logic tree analysis). Risk exposure identification process logic that determines the criticality importance of any single failure mode and selects PM tasks accordingly.

Task. Discrete elements of a work order (sometimes called *scope bullets*) that specifically addresses a single identified equipment failure mode. Generally, a work order comprises many tasks. An overhaul is the extreme example of this observation. Turbine overhauls commonly contain 50 or more discrete tasks on large turbines, all performed at one time possible under a single work order! The task directly relates to the failure mode that must be prevented. Generally, one task addresses a single failure mode. The primary exceptions are high-risk failure modes, such as those involving safety that warrant two or more tasks to ensure a failure mode is clearly defined for condition-directed maintenance performance. Engineers ensure that every critical failure mode has one or more discrete tasks that effectively and applicably address that failure mode.

Task blocking. *See* **Blocking.**

Templates. Standard models built around specific design equipment classes like pumps, motors, or valves (further delineated) that pre-develop and prepare most of the PM information for the equipment. Use in databases (a set of related table records), in spreadsheets, a sheet (worksheet), in document software. A single document comprises templates for reuse modeling plant-installed equipment.

Thermal runaway (in an electronic device). A process whereby heatup increases the current load on the device, increasing the heating of the device, which in turn demands more current. The process is unstable and burns up the transistor, diode, or other electronic device.

Three-Mile Island (TMI). The greatest single U.S. nuclear accident in more than 40 years of commercial power generation.

Time constant. The characteristic time for a transient state change to reach a new value. Mathematically defined as $1/\tau$, where τ is a derived physical constant determined by design or theory (or practically measured by initiating a step change) and measuring time to return to steady conditions.

Train. One of several identical equipment configurations that redundantly provide system output functions.

Travelers. *See* **Riders.**

Tube plugs. Plugs in the plenum or waterbox area of heat exchangers that isolate leaky tubes so the heater can continue to be used.

Tube stakes. Rods inserted into the tubes of eroded tubes to prevent through-wall erosion and failure.

Type (subtype). Equipment component classification by design and sub-design category.

Viton. Trade name for a high-temperature elastomeric compound used for o-rings and seals.

Window weld. A weld made but cutting out a panel *window* and replacing the weak area with a new panel. A more permanent, quality weld technique.

WO. *See* **Work order.**

Work order (WO). An organized scope of work that identifies the work authorizations, requirements, skills, hours estimated, tasks, frequency intervals, tagouts, and tools (among other things) to uniquely specify the work to be done and collect as found (and as left, for that matter) condition details. The fundamental unit of productivity for maintenance shop in an industrial facility application. Completed WOs provide a history of parts usage, failure experience, and maintenance testing validation of work effectiveness and many other details.

The WO is the fundamental document of maintenance performance and can be further classified by scheduled and non-scheduled, calibration, failed equipment, class (ASME code) and Q (nuclear) work, and so on.

Work order scope of work. *See* **Workscopes.**

Workscopes (work order scope of work). The summarized list of tasks and unambiguous instructions that clearly identify their interval and completion.

Yellow dog (slang). Hardcopy scratch notes (from yellowed old paper).

Index

A

B

C

D

E

F

G

H

I–K

L

M

N

O

P

Q

R

S

T

U

V

W–Z